Earth and Sky Every Child Should Know

Easy studies of the earth and the stars for any time and place

Julia Ellen Rogers

Writat

This edition published in 2022

ISBN : 9789356391284

Contents

PART I
THE EARTH

THE GREAT STONE BOOK

"The crust of our earth is a great cemetery where the rocks are tombstones on which the buried dead have written their own epitaphs. They tell us who they were, and when and where they lived."—*Louis Agassiz.*

Deep in the ground, and high and dry on the sides of mountains, belts of limestone and sandstone and slate lie on the ancient granite ribs of the earth. They are the deposits of sand and mud that formed the shores of ancient seas. The limestone is formed of the decayed shells of animal forms that flourished in shallow bays along those shores. And all we know about the life of these early days is read in the epitaphs written on these stone tables.

Under the stratified rocks, the granite foundations tell nothing of life on the earth. But the sea rolled over them, and in it lived a great variety of shellfish. Evidently the earliest fossil-bearing rocks were worn away, for the rocks that now lie on the granite show not the beginnings, but the high tide of life. The "lost interval" of which geologists speak was a time when living forms were few in the sea.

In the muddy bottoms of shallow, quiet bays lie the shells and skeletons of the creatures that live their lives in those waters and die when they grow old and feeble. We have seen the fiddler crabs by thousands on such shores, young and old, lusty and feeble. We have seen the rocks along another coast almost covered by the coiled shells of little gray periwinkles, and big clumps of black mussels hanging on the piers and wharfs. All these creatures die, at length, and their shells accumulate on the shallow sea bottom. Who has not spent hours gathering dead shells which the tide has thrown up on the beach? Who has not cut his foot on the broken shells that lie in the sandy bottom we walk on whenever we go into the surf to swim or bathe?

Read downward from the surface toward the earth's centre—

TABLE OF CONTENTS

Part	Rock Systems		Dominant Animals	Dominant Plants
VII.	Recent		Man	Flowering kinds
	Quaternary			
VI. {		Pliocene	Mammals	Early flowering
	Tertiary	{ Miocene		
		Eocene		
V.	Mesozoic		Reptiles	Cycads
IV.	Carboniferous		Amphibians	Ferns and Conifers
III.	Devonian		Fishes	Ferns
II.	Silurian		Molluscs	Seaweeds
I.	Fire-formed		No life	No life

It is by dying that the creatures of the sea write their epitaphs. The mud or sand swallows them up. In time these submerged banks may be left dry, and become beds of stone. Then some of the skeletons and shells may be revealed in blocks of quarried stone, still perfect in form after lying buried for thousands of years.

The leaves of this great stone book are the layers of rock, laid down under water. Between the leaves are pressed specimens—fossils of animals and plants that have lived on the earth.

THE FOSSIL FISH

I remember seeing a flat piece of stone on a library table, with the skeleton of a fish distinctly raised on one surface. The friend who owned this strange-looking specimen told me that she found it in a stone quarry. She brought home a large piece of the slate, and a stone-mason cut out the block with the fish in it, and her souvenir made a useful and interesting paper-weight.

The story of that fish I heard with wonder, and have never forgotten. I had never heard of fossil animals or plants until my good neighbour talked about them. She showed me bits of stone with fern leaves pressed into them. One piece of hard limestone was as full of little sea-shells as it could possibly be. One ball of marble was a honeycombed pattern, and called "fossil coral."

The fossil fish was once alive, swimming in the sea, and feeding on the things it liked to eat, as all happy fishes do. Near shore a river poured its muddy water into the sea, and the sandy bottom was covered with the mud that settled on it. At last the fish grew old, and perhaps a trifle stupid about catching minnows. It died, and sank to the muddy floor of the sea. Its horny bones were not dissolved by the water. They remained, and the mud filtered in and filled all the spaces. Soon the fish was buried completely by the sediment the river brought.

Years, thousands of them, went by, and the layer of mud was so thick and heavy above the skeleton of the fish that it bore a weight of tons there, under the water. The close-packed mud became a stiff clay. After more thousands of years, the sea no longer came so far ashore, for the river had built up a great delta of land out of mud. The clay in which the fish was hidden hardened into slate. Water crept down in the loose upper layers, dissolving out salt and other minerals, and having harder work to soak through, the lower it went. The water left some of the minerals it had accumulated, calcium and silica and iron, in the lower rock beds, making them harder than they were before, and heavier and less porous.

When the river gorge was cut through these layers of rock, the colour and thickness of each kind were laid bare. Centuries after, perhaps thousands of years, indeed, the quarrymen cut out the layers fit for building stones, flags for walks and slates for roofing. In the splitting of a flagstone, the long-buried skeleton of the fish came to light.

Under our feet the earth lies in layers. Under the soil lie loose beds of clay and sand and gravel, and under these loose kinds of earth are close-packed clays, sandstones, limestones, shales, often strangely tilted away from the horizontal line, but variously fitted, one layer to another. Under these rocks lie the foundations of the earth—the fire-formed rocks, like granite. The depth of this original rock is unknown. It is the substance out of which the earth is made, we think. All the layered rocks are made of particles of the older ones, stolen by wind and water, and finally deposited on the borders of lakes and seas. So our rivers are doing to-day what they have always done—they are tearing down rocks, grinding and sifting the fragments, and letting them fall where the current of fresh water meets a great body of water that is still, or has currents contrary to that of the river.

Do you see a little dead fish in the water? It is on the way to become a fossil, and the mud that sifts over it, to become a layer of slate. Every seashore buries its dead in layers of sand and mud.

THE CRUST OF THE EARTH

It is hard to believe that our solid earth was once a ball of seething liquid, like the red-hot iron that is poured out of the big clay cups into the sand moulds at an iron foundry. But when a mountain like Vesuvius sets up a mighty rumbling, and finally a mass of white-hot lava bursts from the centre and streams down the sides, covering the vineyards and olive orchards, and driving the people out of their homes in terror, it seems as if the earth's crust must be but a thin and frail affair, covering a fiery interior, which might at any time break out. The people who live near volcanoes might easily get this idea.

But they do not. They go back as soon as the lava streams are cooled, and rebuild their homes, and plant more orchards and vineyards. "It is *so* many years," say they to one another, "since the last bad eruption. Vesuvius will probably sleep now till we are dead and gone."

This is good reasoning. There are few active volcanoes left on the earth, compared with the number that were once active, and long ago became extinct. And the time between eruptions of the active ones grows longer; the eruptions less violent. Terrible as were the recent earthquakes of San Francisco and Messina, this form of disturbance of the earth's crust is growing constantly less frequent. The earth is growing cooler as it grows older; the crust thickens and grows stronger as centuries pass. We have been studying the earth only a few hundred years. The crust has been cooling for millions of years, and mountain-making was the result of the shrinking of the crust. That formed folds and clefts, and let masses of the heated substance pour out on the surface.

My first geography lesson I shall never forget. The new teacher had very bright eyes and *such* pretty hands! She held up a red apple, and told us that the earth's substance was melted and burning, inside its crust, which was about as thick, in proportion to the size of the globe, as the skin of the apple. I was filled with wonder and fear. What if we children jumped the rope so hard as to break through the fragile shell, and drop out of sight in a sea of fiery metal, like melted iron? Some of the boys didn't believe it, but they were impressed, nevertheless.

The theory of the heated interior of the earth is still believed, but the idea that flames and bubbling metals are enclosed in the outer layer of solid matter has generally been abandoned. The power that draws all of its particles toward the earth's centre is stated by the laws of gravitation. The amount of "pull" is the measure of the weight of any substance. Lift a stone, and then a feather pillow, much larger than the stone. One is strongly drawn to the earth; the other not. One is *heavy*, we say, the other *light*.

If a stone you can pick up is heavy, how much heavier is a great boulder that it takes a four-horse team to haul. What tremendous weight there is in all the boulders scattered on a hillside! The hill itself could not be made level without digging away thousands of tons of earth. The earth's outer crust, with its miles in depth of mountains and level ground, is a crushing weight lying on the heated under-substance. Every foot of depth adds greatly to the pressure exerted upon the mass, for the attraction of gravitation increases amazingly as the centre of the earth is approached.

It is now believed that the earth is solid to its centre, though heated to a high degree. Terrific pressure, which causes this heat, is exerted by the weight of the crust. A crack in the crust may relieve this pressure at some point, and a mass of substance may be forced out and burst into a flaming stream of lava. Such an eruption is familiar in volcanic regions. The fact that red-hot lava streams from the crater of Vesuvius is no proof that it was seething and bubbling while far below the surface.

Volcanoes, geysers, and hot springs prove that the earth's interior is hot. The crust is frozen the year around in the polar regions, and never between the Tropics of Cancer and Capricorn. The sun's rays produce our different climates, but they affect only the surface. Underground, there is a rise of a degree of temperature for every fifty feet one goes down. The lowest mine shaft is about a mile deep. That is only one four-thousandth of the distance to the earth's centre.

By an easy computation we could locate the known melting-point for metals and other rock materials. But one degree for each fifty feet of depth below the surface may not be correct for the second mile, as it is for the first. Again, the melting-point is probably a great deal higher for substances under great pressure. The weight of the crust is a burden the under-rocks bear. Probably the pressure on every square inch reaches thousands of tons. Could any substance become liquid with such a weight upon it, whatever heat it attained? Nobody can answer this question.

The theory that volcanoes are chimneys connecting lakes of burning lava with the surface of the earth is discredited by geologists. The weight of the overlying crust would, they think, close such chambers, and reduce liquids to a solid condition.

Since the first land rose above the sea, the crust of the earth has gradually become more stable, but even now there is scarcely a day when the instruments called seismographs do not record earthquake shocks in some part of the earth; and the outbreaks of Vesuvius and Ætna, the constant boiling of lava in the craters of the Hawaiian Islands and other volcanic centres, prove that even now the earth's crust is very thin and unstable. The further back in time we go, the thinner was the crust, the more frequent the outbursts of volcanic activity, the more readily did wrinkles form.

The shores of New Jersey and of Greenland are gradually sinking, and the sea coming up over the land. Certain parts of the world are gradually rising out of the sea. In earlier times the rising or the sinking of land over large areas happened much more frequently than now.

WHAT IS THE EARTH MADE OF?

"Baking day" is a great institution in the comfortable farm life of the American people. The big range oven is not allowed to grow cold until rows of pies adorn the pantry shelves, and cakes, tarts, and generous loaves of bread are added to the store. Cookies, perhaps, and a big pan full of crisp, brown doughnuts often crown the day's work. No gallery of art treasures will ever charm the grown-up boys and girls as those pantry shelves charmed the bright-eyed, hungry children, who were allowed to survey the treasure-house, and sample its good things while they were still warm.

You could count a dozen different kinds of cakes and pies, rolls and cookies on those pantry shelves, yet several of them were made out of the same dough. Instead of a loaf of bread, mother could make two or three kinds of coffee cake, or cinnamon rolls, or currant buns, or Parker-House rolls. Even the pastry, which made the pies and tarts, was not so different from the bread dough, for each was made of flour, and contained, besides the salt, "shortening," which was butter or lard. Sugar was used in everything, from the bread, which had a table-spoonful, to the cookies, which were finished with a sifting of sugar on top.

How much of the food we eat is made of a very few staple foodstuffs,—starch, sugar, fats! So in the wonderful earth and all that grows out of it and lives upon it. Only seventy different elements have been discovered, counting, besides the earth, the water and the air, and even the strange wandering bodies, called meteorites, that fall upon the earth out of the sky. Like the flour in the different cakes and pies, the element carbon is found in abundance and in strangely different combinations. As a gas, in combination with oxygen, it is breathed out of our lungs, and out of chimneys where coal and wood are burned. It forms a large part of the framework of trees and other plants, and remains as charcoal when the wood is slowly burned under a close covering. There is a good proportion of carbon in animal bodies, in the bones as well as the soft parts, and carbon is plentiful in the mineral substances of the earth.

The chemist is the man who has determined for us the existence and the distribution of the seventy elements. He finds them in the solid substances of the globe and in the water that covers four-fifths of its surface; in the atmosphere that covers sea and land, and in all the living forms of plants and animals that live in the seas and on the land. By means of an instrument called the spectroscope, the heavenly bodies are proved to be made of the same substances that are found in the rocks. The sun tells what it is made of, and one proof that the earth is a child of the sun is in the fact that the same elements are found in the substance of both.

Of the seventy elements, the most important are these: Oxygen, silicon, aluminum, iron, manganese, calcium, magnesium, potassium, sodium, carbon, hydrogen, phosphorus, sulphur, chlorine, nitrogen.

Oxygen is the most plentiful and the most important element. One-fifth of the air we breathe is oxygen; one-third of the water we drink. The rock foundations of the earth are nearly one-half oxygen. No fire can burn, no plant or animal can grow, or even decay after it dies, unless oxygen is present and takes an active part in each process. Strangely enough, this wonderful element is invisible. We open a window, and pure air, rich in oxygen, comes in and takes the place of the bad air but we cannot see the change. Water we see, but if the oxygen and the hydrogen which compose the colourless liquid were separated, each would become at once an invisible gas. The oxygen of solid rocks exists only in combination with other elements.

Silicon is the element which, united with oxygen, makes the rock called quartz. On the seashore the children are busy with their pails and shovels digging in the white, clean sand. These grains are of quartz,—fine crystals of a rock which forms nearly three-quarters of the solid earth's substance. Not only in rocks, but out here in the garden, the soil is full of particles of sand. You cannot get away from it.

Aluminum is a light, bluish-white metal which we know best in expensive cooking utensils. It is more abundant even than iron, but processes of extracting it from the clay are still expensive. It is oftenest found in combination with oxygen and silicon. While nearly one-tenth of the earth's crust is composed of the metal aluminum, four-fifths and more is composed of the minerals called silicates of aluminum—oxygen, silicon, and aluminum in various combinations. It is more plentiful than any other substance in rocks and in the clays and ordinary soils, which are the finely ground particles of rock material.

Iron is one of the commonest of elements. We know it by its red colour. A rusty nail is covered with oxide of iron, a combination which is readily formed wherever iron is exposed to the action of water or air. You have seen yellowish or red streaks in clefts of the rocks. This shows where water has dissolved out the iron and formed the oxide. The red colour of New Jersey soil is due to the iron it contains. Indeed, the whole earth's crust is rich in iron which the water easily dissolves. The roots of plants take up quantities of iron in solution and this mounts to the blossoms, leaves, and fruit. The red or yellow colour of autumn leaves, of apples, of strawberries, of tulips, and of roses, is produced by iron. The rosy cheeks of children are due to iron in the food they eat and in the water they drink. The doctor but follows the suggestion of nature when he gives a pale and listless person a tonic of iron to make his blood red.

Iron is rarely found free, but it forms about five per cent. of the crust of the earth, and it is believed to form at least one-fifth of the unknown centre of the earth, the bulk of the globe, the weight of which we know, but concerning the substance of which we can say little that is positive.

Manganese is not a conspicuous element, but is found united with oxygen in purplish or black streaks on the sides of rocks. It is somewhat like iron, but much less common.

Calcium is the element that is the foundation of limestones. The skeletons and shells of animals are made of calcite, a common mineral formed by the uniting of carbon, oxygen, and calcium. Marbles are, perhaps, the most permanent form of the limestone rocks. "Hard" water has filtered through rocks containing calcite, and absorbed particles of this mineral. From water thus impregnated, all animal life on the earth obtains its bone-building and shell-building materials.

Carbon forms a large part of the tissues of plants and animals, and in the remains of these it is chiefly found in the earth's crust. When these burn or decay, the carbon remains as charcoal or escapes to the air in union with oxygen as the well known carbonic acid gas. This is one of the most important foods of plants. Joined with calcium it forms the mineral calcite, or carbonate of lime.

Hydrogen is one of the two gases that unite to form water. Oxygen is the other. Many kinds of rock contain a considerable amount of water. Surface water sinks into porous soils and rocks, and accumulates in pockets and veins which feed springs, and are the reserve water supply that keeps our rivers flowing, even through dry weather. More water is held by absorption in the earth's solid crust than in all the oceans and seas and great lakes.

Hydrogen, combined with carbon, occurs in solid rocks where the remains of plants and animals have slowly decayed. From such processes the so-called hydrocarbons, rock oil and natural gas, have accumulated. When such decay goes on above ground, these valuable products escape into the air. Marsh gas, whose feeble flame above decaying vegetation is the will-o'-the wisp of swamps, is an example.

Magnesium, *potassium*, and *sodium* are found in equal quantities in the earth's crust, but never free. In union with chlorine, each forms a soluble salt, and is thus found in water. Common salt, chloride of sodium, is the most abundant of these. Water dissolves salt out of the rocks, and carries it into the sea. Clouds that rise by the evaporation of ocean water leave the salt behind, hence the seas are becoming more and more salty, for the rivers carry salt to the oceans, which hold fast all they get.

Phosphorus is an element found united with oxygen in the tissues of both plants and animals. It is most abundant in bones. Rocks containing fossil bones are rich in lime phosphates, which are important commercial fertilizers for enriching the soil. Beds of these rocks are found and mined in South Carolina and elsewhere.

Sulphur is well known as a yellow powder found most plentifully in rocks that are near volcanoes. It is a needed element in plant and animal bodies. It occurs in rocks, united with many different elements. In union with oxygen and a metal it forms the group of minerals called sulphates. In union with iron it forms sulphide of iron. The "fool's gold" which Captain John Smith's colonists found in the sand at Jamestown, was this worthless iron pyrites.

Chlorine is a greenish, yellow gas, very heavy, and dangerous to inhale. If it gets into the lungs, it settles into the lowest levels, and one must stand on one's head to get it out. As an element of the earth's crust it is not very plentiful, but it is a part of all the chlorides of sodium, magnesium, and potassium. In salt, it forms two per cent. of the sea water. It is much less abundant in the rocks.

To these elements we might add *nitrogen*, that invisible gas which forms nearly four-fifths of our atmosphere, and is a most important element of plant food in the soil. Most of the seventy elements are very rare. Many are metals, like gold and iron and silver. Some are not metals. Some are solid. A few are liquid, like the metal mercury, and several are gaseous. Some are free and pure, and show no disposition to unite with others. Nuggets of gold are examples of this. Some exist only in union with other elements. This is the common rule among the elements. Changes are constantly going on. The elements are constantly abandoning old partnerships and forming new ones. Growth and decay of plant and animal life are but parts of the great programme of constant change which is going on and has been in progress since the world began.

THE FIRST DRY LAND

When the earth's crust first formed it was still hot, though not so hot as when it was a mass of melted, glowing substance. As it moved through the cold spaces of the sky, it lost more heat and its crust became thicker. At length the cloud masses became condensed enough to fall in torrents of water, and a great sea covered all the land. This was before any living thing, plant or animal, existed on our planet. Can you imagine the continents and islands that form the land part of a map or globe suddenly overwhelmed by the oceans, the names and boundaries of which you have taken such pains to learn in the study of geography? The globe would be one blank of blue water, and geography would be abolished—and there would be nobody to study it. Possibly the fishes in the sea might not notice any change in the course of their lives, except when they swam among the ruins of buried cities and peered into the windows of high buildings, or wondered what new kind of seaweed it was when they came upon a submerged forest.

In that old time of the great sea that covered the globe, we are told that there was a dense atmosphere over the face of the deep. So things were shaping themselves for the far-off time when life should exist, not only in the sea, where the first life did appear, but on land. But it took millions of years to fit the earth for living things.

The cooling of the earth made it shrink, and the crust began to be folded into gentle curves, as the skin of a shrunken apple becomes wrinkled on the flesh. Some of these creases merely changed the depth of water on the sea bottom; but one ridge was lifted above the water. The water parted and streamed down its sloping sides, and a granite reef, which shone in the sunshine, became the first dry land. It lay east and west, and stretched for many miles. It is still dry land and is a part of our own continent. Now it is but a small part of the country, but it is known by geologists, who can tell its boundaries, though newer land joins it on every side. It is named the Laurentian Hills, on geological maps. Its southern border reaches along the northern boundary of the Great Lakes to the head-waters of the Mississippi River.

From this base, two ridges are lifted, forming a colossal V. One extends northeast to Nova Scotia; the other northwest to the Arctic seas. The V encloses Hudson Bay.

Besides this first elongated island of bare rocks, land appeared in a strip where now the Blue Ridge Mountains stretch from New England to Georgia. The other side of the continent lifted up two folds of the crust above sea level. They are the main ridges of the Colorado and the Wasatch Mountains.

Possibly the main ridge of the Sierra Nevada rose also at this time. The Ozark group of mountains, too, showed as a few island peaks above the sea.

These first rocks were rapidly eaten away, for the atmosphere was not like ours, but heavily charged with destructive gases, which did more, we believe, to disintegrate the exposed rock surfaces than did the two other forces, wind and water, combined. The sediment washed down to the sea by rains, accumulated along the shores, filling the shallows and thus adding to the width of the land areas. The ancient granite ridge of the Laurentian Hills is now low, and slopes gently. This is true of all very old mountains. The newer ones are high and steep. It takes time to grind down the peaks and carry off the waste material loosened by erosion.

Far more material than could have been washed down the slopes of the first land ridges came directly from the interior of the earth, and spread out in vast, submarine layers upon the early crust. Volcanic craters opened under water, and poured out liquid mineral matter, that flowed over the sea bottom before it cooled. Imagine the commotion that agitated the water as these submerged chimneys blew off their lids, and discharged their fiery contents! It was long before the sea was cool enough to be the home of living things.

The layers of rock that formed under the sea during this period of the earth's history are of enormous thickness. They were four or five miles deep along the Laurentian Hills. They broadened the original granite ridge by filling the sea bottom along the shores. The backbones of the Appalachian system and the Cordilleras were built up in the same way—the oldest rocks were worn away, and their débris built up newer ones in strata.

When these layers of rock became dry land, the earth's crust was much more stable and cool than it had ever been before. The vast rock-building of that era equals all that has been done since. The layers of rocks formed since then do not equal the total thickness of these first strata. So we believe that the time required to build those Archæan rock foundations equals or surpasses the vast period that has elapsed since the Archæan strata were formed.

The northern part of North America has grown around those old granite ridges by the gradual rising of the shores. The geologist may walk along the Laurentian Hills, that parted the waters into a northern and a southern ocean. He crosses the rocky beds deposited upon the granite; then the successive beds formed as the land rose and the ocean receded. Age after age is recorded in the rocks. Gradually the sea is crowded back, and the land masses, east, west, and north, meet to form the continent. Nowhere on the earth are the steps of continental growth shown in unbroken sequence as they are in North America.

How long ago did those first islands appear above the sea? Nobody ventures a definite answer to this question. No one has the means of knowing. But those who know most about it estimate that at the least one hundred million years have passed since then—one hundred thousand thousand years!

A STUDY OF GRANITE

In Every village cemetery it is easy to find shafts of gray or speckled granite, the polished surfaces of which show that the granite is made of small bits of different coloured minerals, cemented together into solid rock. Outside the gate you will usually find a place where monuments and gravestones may be bought. Here there is usually a stonecutter chipping away on a block with his graving tools. He is a man worth knowing, and because his work is rather monotonous he will probably be glad to talk to a chance visitor and answer questions about the different kinds of stone on which he works.

There are bits of granite lying about on the ground. If you have a hand-glass of low power, such as the botany class uses to examine the parts of flowers, it will be interesting to look through it and see the magnified surface of a flake of broken granite. Here are bits of glassy quartz, clear and sparkling in the sun. Black and white may be all the colours you make out in this specimen, or it may be that you see specks of pink, dark green, gray, and smoky brown, all cemented together with no spaces that are not filled. The particles of quartz are of various colours, and are very hard. They scratch glass, and you cannot scratch them with the steel point of your knife, as you can scratch the other minerals associated with the grains of quartz.

Granite is made of quartz, feldspar, and mica, sometimes with added particles of hornblende. Feldspar particles have as wide a range of colour as quartz, but it is easy to tell the two apart. A knife will scratch feldspar, as it is not so hard as quartz. The crystals of feldspar have smooth faces, while quartz breaks with a rough surface as glass does. Feldspar loses its glassy lustre when exposed to the weather, and becomes dull, with the soft lustre of pearl.

Mica may be clear and glassy, and it ranges in colour from transparency through various shades of brown to black. It has the peculiarity of splitting into thin, leaf-like, flexible sheets, so it is easy to find out which particles in a piece of granite are mica. One has only to use one's pocket knife with a little care. Hornblende is a dark mineral which contains considerable iron. It is found in lavas and granites, where it easily decays by the rusting of the iron. It is not unusual to see a rough granite boulder streaked with dark red rust from this cause.

The crumbling of granite is constantly going on as a result of the exposure of its four mineral elements to the air. Quartz is the most stable and resistant to weathering. Soil water trickling over a granite cliff has little effect on the quartz particles; but it dissolves out some of the silicon. The bits of feldspar are even more resistant to water than quartz is, but the air causes them to decay rapidly, and finally to fall away in a sort of mealy clay. Mica, like feldspar, decays easily. Its substance is dissolved by water and carried away to become a kind of clay. The hornblende rusts away chiefly under the influence of moist air and trickling water.

We think of granite as a firm, imperishable kind of rock, and use it in great buildings like churches and cathedrals that are to stand for centuries. But the faces that are exposed to the air suffer, especially in regions having a moist climate. The signs of decay are plainly visible on the outer surfaces of these stones. Fortunate it is that the weathering process cannot go very deep.

The glassy polish on a smooth granite shaft is the silicon which acts as a cement to bind all the particles together. It is resistant to the weather. A polished shaft will last longer than an unpolished one.

Granites differ in colouring because the minerals that compose them, the feldspars, quartzes, micas, and hornblendes, have each so wide a range of colour. Again, the proportions of the different mineral elements vary greatly in different granites. A banded granite the colours of which give it a stratified appearance is called a gneiss.

We have spoken before of the seventy elements found in the earth's crust. A mineral is a union of two or more of these different elements; and we have found four minerals composing our granite rock. It may be interesting to go back and inquire what elements compose these four minerals. Quartz is made of silicon and oxygen. Feldspar is made of silicon, oxygen, and aluminum. Mica is made of silicon, oxygen, and carbon, with some mingling of potassium and iron and other elements in differing proportions. Hornblende is made of silicon, oxygen, carbon, and iron.

The crumbling of a granite rock separates the minerals that compose it, reducing some to the condition of clay, others to grains of sand. Some of the elements let go their union and become free to form new unions. Water and wind gather up the fragments of crumbling granite and carry them away. The feldspar and mica fragments form clay; the quartz fragments, sand. All of the sandstones and slates, the sand-banks and sand beaches, are made out of crumbled granite, the rocky foundations of the earth.

METAMORPHIC ROCKS

In the dawn of life on the earth, soft-bodied creatures, lowest in the scale of being, inhabited the sea. The ancient volcanoes the subterranean eruptions of which had spread layers of mineral substance on the ocean floor, and heated the water to a high degree, had subsided. The ocean was sufficiently cool to maintain life. The land was being worn down, and its débris washed into the ocean. The first sand-banks were accumulating along sandy shores. The finer sediment was carried farther out and deposited as mud-banks. These were buried under later deposits, and finally, by the rising of the earth's crust, they became dry land. Time and pressure converted the sand-banks into sandstones; the mud-banks into clay. The remains of living creatures utterly disappeared, for they had no hard parts to be preserved as fossils.

The shrinking of the earth's crust had crumpled into folds of the utmost complexity those horizontal layers of lava rock poured out on the ocean floor. Next, the same forces attacked the thick rock layers formed out of sediment—the aqueous or water-formed sandstones and clays.

The core of the globe contracts, and the force that crumples the crust to fit the core generates heat. The alkaline water in the rocks joins with the heat produced by the crumpling and crushing forces, acting downward, and from the sides, to transform pure sandstone into glassy quartzite, and clay into slate. In other words, water-formed rocks are baked until they become fire-formed rocks. They are what the geologist calls *metamorphic*, which means *changed.*

In many mountainous regions there are breaks through the strata of sandstone and slates and limestones, through which streams of lava have poured forth from the heated interior. Along the sides of these fissures the hot lava has changed all the rocks it touched. The heat of the volcanic rock matter has melted the silica in the sand, which has hardened again into a crystalline substance like glass.

Have you ever visited a brick-yard? Here men are sifting clay dug out of a pit or the side of a hill, adding sand from a sand-bank, and in a big mixing box, stirring these two "dry ingredients" with water into a thick paste. This dough is moulded into bricks, sun-dried, and then baked in kilns themselves built of bricks. At the end of the baking, the soft, doughy clay block is transformed into a hard, glassy, or dull brick. From aqueous rock materials, fire has produced a metamorphic rock. Volcanic action is imitated in this common, simple process of brickmaking.

Milwaukee brick is made of clay that has no iron in it. For this reason the bricks are yellow after baking. Most bricks are red, on account of the iron in the clay, which is converted into a red oxide, or rust, by water and heat.

Common flower pots and the tiles used in draining wet land are not glazed, as hard-burned bricks are. The baking of these clay things is done with much less heat. They are left somewhat porous. But the tiles of roofs are baked harder, and get a surface glaze by the melting of the glassy particles of the sand.

As bricks vary in colour and quality according to the materials that compose them, so the metamorphic rocks differ. The white sand one sees on many beaches is largely quartz. This is the substance of pure, white sandstone. Metamorphism melts the silica into a glassy liquid cement; the particles are bound close together on cooling. The rock becomes a white, granular quartzite, that looks like loaf sugar. If banded, it is called gneiss. Such rocks take a fine polish.

Pure limestone is also white and granular. When metamorphosed by heat, it becomes white marble. The glassy cement that holds the particles of lime carbonate shows as the glaze of the polished surface. It is silica. One sees the same mineral on the face of polished granite.

Clays are rarely pure. Kaolin is a white clay which, when baked, becomes porcelain. China-ware is artificially metamorphosed kaolin. In the early rocks the clay beds were transformed by heat into jasper and slates. In beds where clay mingled with sand, in layers, gneiss was formed. If mica is a prominent element, the metamorphic rock is easily parted into overlapping, scaly layers. It is a mica schist. If hornblende is the most abundant mineral, the same scaly structure shows in a dark rock called hornblende schist, rich in iron. A schist containing much magnesia is called serpentine.

The bricks of the wall, the tiles on the roof, the flower pots on the window sill, and the dishes on the breakfast table, are examples of metamorphic rocks made by man's skill, by the use of fire and water acting on sand and clay. Pottery has preserved the record of civilization, from the making of the first crude utensils by cave men to the finest expression of decorative art in glass and porcelain.

The choicest material of the builder and the sculptor is limestone baked by the fires under the earth's crust into marble. The most enduring of all the rocks are the foundation granite, and the metamorphic rocks that lie next to them. Over these lie thick layers of sedimentary rocks laid down by water. In them the record of life on the earth is written in fossils.

THE AIR IN MOTION

Most of the beautiful things that surround us and make our lives full of happiness appeal to one or more of our five senses. The green trees we can see, the bird songs we hear, the perfume of honey-laden flowers we smell, the velvety smoothness of a peach we feel, and its rich pulp we taste. But over all and through all the things we see and feel and hear and taste and smell, is the life-giving air, that lies like a blanket, miles in depth, upon the earth. The substance which makes the life of plants and animals possible is, when motionless, an invisible, tasteless, odourless substance, which makes no sound and is not perceptible to the touch.

Air fills the porous substance of the earth's crust for a considerable distance, and even the water has so much air in it that fishes are able to breathe without coming to the surface. It is not a simple element, like gold, or carbon, or calcium, but is made up of several elements, chief among which are nitrogen and oxygen. Four-fifths of its bulk is nitrogen and one-fifth oxygen. There is present in air more or less of watery vapour and of carbon dioxide, the gas which results from the burning or decay of any substance. Although no more than one per cent. of the air that surrounds us is water, yet this is a most important element. It forms the clouds that bear water back from the ocean and scatter it in rain upon the thirsty land. Solid matter in the form of dust, and soot from chimneys, accumulates in the clouds and does a good work in condensing the moisture and causing it to fall.

It is believed that the air reaches to a height of one hundred to two hundred miles above the earth's surface. If a globe six feet in diameter were furnished with an atmosphere proportionately as deep as ours, it would be about an inch in depth. At the level of the sea the air reaches its greatest density. Two miles above sea-level it is only two-thirds as dense. On the tops of high mountains, four or five miles above sea level, the air is so rarefied as to cause the blood to start from the nostrils and eyelids of explorers. The walls of the little blood-vessels are broken by the expansion of the air that is inside. At the sea-level air presses at the rate of fifteen pounds per square foot in all directions. As one ascends to higher levels, the air pressure becomes less and less.

The barometer is the instrument by which the pressure of air is measured. A glass tube, closed at one end, and filled with mercury, the liquid metal often called quicksilver, is inverted in a cup of the same metal, and so supported that the metal is free to flow between the two vessels. The pressure of air on the surface of the mercury in the cup is sufficient at the sea-level to sustain a column of mercury thirty inches high in the tube. As the instrument is carried up the side of a mountain the mercury falls in the tube. This is because the air pressure decreases the higher up we go. If we should descend into the shaft of the deepest mine that reaches below the sea level, the column of air supported by the mercury in the cup would be a mile higher, and for this reason its weight would be correspondingly greater. The mercury would thus be forced higher in the tube than the thirty-inch mark, which indicates sea-level.

Another form of barometer often seen is a tube, the lower and open end of which forms a U-shaped curve. In this open end the downward pressure of the air rests upon the mercury and holds it up in the closed end, forcing it higher as the instrument is carried to loftier altitudes. At sea level a change of 900 feet in altitude makes a change of an inch in the height of the mercury in the column. The glass tube is marked with the fractions of inches, or of the metre if the metric system of measurements is used.

It is a peculiarity of air to become heated when it is compressed, and cooled when it is allowed to expand again. It is also true that when the sun rises, the atmosphere is warmed by its rays. This is why the hottest part of the day is near noon when the sun's rays fall vertically. The earth absorbs a great deal of the sun's heat in the daytime and through the summer season. When it cools this heat is given off, thus warming the surrounding atmosphere. In the polar regions, north and south, the air is far below freezing point the year round. In the region of the Equator it rarely falls below 90 degrees, a temperature which we find very uncomfortable, especially when there is a good deal of moisture in the air.

If we climb a mountain in Mexico, we leave the sultry valley, where the heat is almost unbearable, and very soon notice a change. For every three hundred feet of altitude we gain there is a fall of one degree in the temperature. Before we are half way up the slope we have left behind the tropical vegetation, and come into a temperate zone, where the plants are entirely different from those in the lower valley. As we climb, the vegetation becomes stunted, and the thermometer drops still lower. At last we come to the region of perpetual snow, where the climate is like that of the frozen north.

So we see that the air becomes gradually colder as we go north or south from the Equator, and the same change is met as we rise higher and higher from the level of the sea.

It is only when air is in motion that we can feel and hear it, and there are very few moments of the day, and days of the year, when there is not a breeze. On a still day fanning sets the air in motion, and creates a miniature breeze, the sound of which we hear in the swishing of the fan. The great blanket of air that covers the earth is in a state of almost constant disturbance, because of the lightness of warm air and the heaviness of cold air. These two different bodies are constantly changing places. For instance, the heated air at the Equator is constantly being crowded upward by cold air which settles to the level of the earth. Cold streams of air flow to the Tropics from north and south of the Equator, and push upward the air heated by the sun.

This constant inrush of air from north and south forms a double belt of constant winds. If the earth stood still, no doubt the direction would be due north and due south for these winds; but the earth rotates rapidly from west to east upon its axis, carrying with it everything that is securely fastened to the surface: the trees, the houses, etc. But the air is not a part of the earth, not even so much as the seas, the waters of which must stay in their proper basins, and be whirled around with other fixed objects. The earth whirls so rapidly that the winds from north and south of the Equator lag behind, and thus take a constantly diagonal direction. Instead of due south the northern belt of cold air drifts south-west and the southern belt drifts northwest. These are called the Trade Winds. Near the Equator they are practically east winds.

The belt of trade winds is about fifty degrees wide. It swings northward in our summer and southward in our winter, its centre following the vertical position of the sun. Near the centre of the course which marks the meeting of the northern with the southern winds is a "Belt of Calms" where the air draws upward in a strong draught. The colder air of the trade winds is pushing up the columns of light, heated air. This strip is known by sailors as "the Doldrums," or "the region of equatorial calms." Though never wider than two or three hundred miles, this is a region dreaded by captains of sailing-vessels, for they often lie becalmed for weeks in an effort to reach the friendly trade winds that help them to their desired ports. Vessels becalmed are at the mercy of sudden tempests which come suddenly like thunder-storms, and sometimes do great damage to vessels because they take the sailors unawares and allow no time to shorten sail.

Until late years the routes of vessels were charted so that sailors could take advantage of the trade winds in their long voyages. It was necessary in the days of sailing-vessels for the captain to understand the movements of winds which furnished the motive power that carried his vessel. Fortunate it was for him that there were steady winds in the temperate zones that he could take advantage of in latitudes north of the Tropic of Cancer and south of the Tropic of Capricorn. What becomes of the hot air that rises in a constant stream above the "Doldrums," pushed up by the cooler trade winds that blow in from north and south? Naturally this air cannot ascend very high, for it soon reaches an altitude in which its heat is rapidly lost, and it would sink if it were not constantly being pushed by the rising column of warm air under it. So it turns and flows north and south at a level above the trade winds. Not far north of the Tropic of Cancer it sinks to the level of sea and land, and forms a belt of winds that blows ships in a northeasterly direction. Between trades and anti-trades is another zone of calms,—near the Tropics of Cancer and of Capricorn.

The land masses of the continents with their high mountain ranges interfere with these winds, especially in the northern hemisphere, but in the Southern Pacific and on the opposite side of the globe the "Roaring Forties," as these prevailing westerly winds are known by the sailors, have an almost unbroken waste of seas over which they blow. In the long voyages between England and Australia, and in the Indian trade, the ships of England set their sails to catch the roaring forties both going and coming. They accomplish this by sailing past the Cape of Good Hope on the outward voyage and coming home by way of Cape Horn, thus circling the globe with every trip. In the North Atlantic, traffic is now mostly carried on in vessels driven by engines, not by sails. Yet the westerly winds that blow from the West Indies diagonally across the Atlantic are still useful to all sailing craft that are making for British ports.

From the north and from the south cold air flows down into the regions of warmer climate. These polar winds are not so important to sea commerce, but they do a great work in tempering the heat in the equatorial regions. We cannot know how much our summers are tempered by the cool breath of winds that blow over polar ice-fields. And the cold regions of the earth, in their brief summer, enjoy the benefits of the warm breezes that flow north and south from the heated equatorial regions.

The land, north and south, is made habitable by the clouds. They gather their burdens of vapour from the warm seas, the wind drifts them north and south, where they let it fall in rains that make and keep the earth green and beautiful. From the clouds the earth gathers, like a great sponge, the water that stores the springs and feeds rivers and lakes. How necessary are the winds that transport the cloud masses!

The air is the breath of life to all living things on our planet. Mars is one of the sun's family so provided. Plants or animals could probably live on the planet Mars. Do we think often enough of this invisible, life-giving element upon which we depend so constantly?

The open air which the wind purifies by keeping it in motion is the best place in which to work, to play, and to sleep, when work and play are done and we rest until another day comes. Indoors we need all the air we can coax to come in through windows and doors. Fresh air purifies air that is stale and unwholesome from being shut up. Nobody is afraid, nowadays, to breathe night air! What a foolish notion it was that led people to close their bedroom windows at night. Clean air, in plenty, day and night, we need. Air and sunshine are the two best gifts of God.

THE WORK OF THE WIND

When the March wind comes blustering down the street, rudely dashing a cloud of dust in our faces, we are uncomfortable and out of patience. We duck our heads and cover our faces, but even then we are likely to get a cinder in one eye, to swallow germs by the dozens, and to get a gray coating of plain, harmless dust. We welcome the rain that lays the dust, or its feeble imitation, the water sprinkler, that brings us temporary relief.

On the quietest day, even after a thorough sweeping and dusting of the library, you are able to write your name plainly on the film of dust that lies on the polished table. Take a book from the open shelves, and blow into the trough of its top. This is always dusty. Where does the dust come from? This is the house-keeper's riddle.

The answer is not a hard one. I look out of my window on a street which is famous as the road Washington took on his retreat from White Plains to Trenton. It has always been the main thoroughfare between New York and Philadelphia, and now is the route that automobiles follow. A constant procession of vehicles passes my house, and to-day each one approaches in a cloud of dust. The air is gray with suspended particles of dirt. The wind carries the successive clouds, and they roll up against the houses like breakers on the beach. Windows and doors are loose enough to let dust sift in. When a door opens, the cloud enters and lights on rugs and carpets and curtains. Any ledge collects its share of dust. The beating of carpets and rugs disturbs the accumulated dust of many months.

In this lonely Arizona desert the wind drifts the sand into dunes, just as it does on the toe of Cape Cod

The Grand Canyon of the Colorado shows on a magnificent scale the work of water in cutting away rock walls

The wind sweeps the ploughed field, and takes all the dust it can carry. It blows the finest top soil from our gardens into the street. It blows soil from other fields and gardens into ours, so the level of our land is not noticeably lowered. The wind strips the high land and drops its burden on lower levels. This is one of the big jobs the water has to do, and the wind is a valuable helper. To tear down the mountains and fill in the valleys is the great work of the two partners, wind and water.

Dead, still air holds the finest dust, without letting it fall. The buoyancy of the particles overcomes their weight. We see them in a sunbeam, like shining points of precious metal, and watch them. A light breeze picks up bits of soil and litter, from the smallest up to a certain size and weight. If the velocity of the wind increases, its carrying power increases. It is able to carry bits that are larger and heavier. The following table is exact and interesting:

	Velocity in Miles per Hour	*Pressure in Pounds per Sq. Ft.*
Light breeze	14	1
Strong breeze	42	9
Strong gale	70	25
Hurricane	84	36

The terrible paths of hurricanes are seen in forest countries. The trees are uprooted, as if a great roller had crushed them, throwing the tops all in one direction, and leaving the roots uncovered, and a sunken pocket where each tree stood. On a steep, rocky slope, the uprooting of scattered trees often loosens tons of rock, and sends the mass thundering down the mountain-side. Much more destruction may be accomplished by one brief tornado than by years of wear by ordinary breezes.

The wind does much to help the waves in their patient beating on rocky shores. If the wind blows from the ocean and the tide is landward, the two forces combine, and the loose rocks are thrown against the solid beach with astonishing force. Even the gravel and the sharp sand are tools of great usefulness to the waves in grinding down the resisting shore. Up and back they are swept by the water, and going and coming they have their chance to scratch or strike a blow. Boulders on the beach become pockmarked by the constant sand-blast that plays upon them. The lower windows of exposed seaside houses are dimmed by the sand that picks away the smooth surface outside, making it ground glass by the same process used in the factory. Lighthouses have this difficulty in keeping their windows clear. The "lantern" itself is sometimes reached by the sand grains. That is the cupola in which burns the great light that warns vessels away from the rocks and tells the captain where he is.

In the Far Western States the telegraph poles and fence posts are soon cut off at the ground by the flinty knives the wind carries. These are the grains of sand that are blown along just above the ground. The trees are killed by having their bark girdled in this way. The sand-storms which in the orange and lemon region of California are called "Santa Anas" sometimes last two or three days, and damage the trees by piercing the tender bark with the needle-pointed sand.

Wind-driven soil, gathered from the sides of bare hills and mountains, fills many valleys of China with a fine, hard-packed material called "loess." In some places it is hundreds of feet deep. The people dig into the side of a hill of this loess and carry out the diggings, making themselves homes, of many rooms, with windows, doors, and solid walls and floors, all in one solid piece, like the chambered house a mole makes underground in the middle of a field. So compact is the loess that there is no danger of a cave-in.

The hills of sand piled up on the southern shore of Lake Michigan, and at Provincetown, at the toe of Cape Cod, are the work of the wind. On almost any sandy shore these "dunes" are common. The long slope is toward the beach that furnishes the sand. The wind does the building. Up the slope it climbs, then drops its burden, which slides to the bottom of an abrupt landward steep. There is a gradual movement inland if the strongest winds come from the water. The shifting of the dunes threatens to cover fertile land near them. In the desert regions, the border-land is always in danger of being taken back again, even though it has been reclaimed from the desert and cultivated for long years.

Besides tearing down, carrying away, and building up again the fragments of the earth's crust, the wind does much that makes the earth a pleasant planet to live on. It drives the clouds over the land, bringing rains and snows and scattering them where they will bless the thirsty ground and feed the springs and brooks and rivers. It scatters the seeds of plants, and thus plants forests and prairies and lovely mountain slopes, making the wonderful wild gardens that men find when they first enter and explore a new region. The trade winds blow the warm air of the Tropics north and south, making the climate of the northern countries milder than it would otherwise be. Sea winds blow coolness over the land in summer, and cool lake breezes temper the inland regions. From the snow-capped mountains come the winds that refresh the hot, tired worker in the valleys.

Everywhere the wind blows, the life-giving oxygen is carried. This is what we mean when we speak of fresh air. Stagnant air is as unwholesome as stagnant water. Constant moving purifies both. So we must give the wind credit for some of the greatest blessings that come into our lives. Light and warmth come from the sun. Pure water and pure air are gifts the bountiful earth provides. Without them there would be no life on the earth.

RAIN IN SUMMER

How beautiful is the rain!After the dust and heat,In the broad and fiery street,In the narrow lane,How beautiful is the rain!

How it clatters along roofs,Like the tramp of hoofs!How it gushes and struggles outFrom the throat of the overflowing spout!Across the window-pane

It pours and pours;And swift and wide,With a muddy tide,Like a river down the gutter roarsThe rain, the welcome rain!

The sick man from his chamber looksAt the twisted brooks;He can feel the coolBreath of each little pool;His fevered brainGrows calm again,And he breathes a blessing on the rain.

—HENRY W. LONGFELLOW.

WHAT BECOMES OF THE RAIN?

The clouds that sail overhead are made of watery vapour. Sometimes they look like great masses of cotton-wool against the intense blue of the sky. Sometimes they are set like fleecy plumes high above the earth. Sometimes they hang like a sullen blanket of gray smoke, so low they almost touch the roofs of the houses. Indeed, they often rest on the ground and then we walk through a dense fog.

In their various forms, clouds are like wet sponges, and when they are wrung dry they disappear—all their moisture falls upon the earth. When the air is warm, the water comes in the form of rain. If it is cold, the drops are frozen into hail, sleet, or snow.

All of the water in the oceans, in the lakes and rivers, great and small, all over the earth, comes from one source, the clouds. In the course of a year enough rain and snow fall to cover the entire surface of the globe to a depth of forty inches. This quantity of water amounts to 34,480 barrels on every acre. What becomes of it all?

We can easily understand that all the seas and the other bodies of water would simply add forty inches to their depth, and many would become larger, because the water would creep up on their gradually sloping shores. We have to account for the rain and the snow that fall upon the dry land and disappear.

Go out after a drenching rainstorm and look for the answer to this question. The gullies along the street are full of muddy, running water. There are pools of standing water on level places, but on every slope the water is hurrying away. The ground is so sticky that wagons on country roads may mire to the hubs in the pasty earth. There is no use in trying to work in the garden or to mow the lawn. The sod is soft as a cushion, and the garden soil is water-soaked below the depth of a spading-fork.

The sun comes out, warm and bright, and the flagstones of the sidewalk soon begin to steam like the wooden planks of the board walk. The sun is changing the surface water into steam which rises into the sky to form a part of another bank of clouds. The earth has soaked up quantities of the water that fell. If we followed the racing currents in the gullies we should find them pouring into sewer mains at various points, and from these underground pipes the water is conducted to some outlet like a river. All of the streams are swollen by the hundreds of brooks and rivulets that are carrying the surface water to the lowest level.

Rain and wind are the sculptors that have carved these strange castles out of a rocky table

All the water in the seas, lakes, rivers, and springs came out of the clouds

So we can see some of the rainfall going back to the sky, some running off through rivulets to the sea, and some soaking into the ground. It will be interesting to follow this last portion as it gradually settles into the earth. The soil will hold a certain quantity, for it is made up of fine particles, all separated by air spaces, and it acts like a sponge. In seasons of drought and great heat the sun will draw this soil water back to the surface, by forming cracks in the earth, and fine, hair-like tubes, through which the vapour may easily rise. The gardener has to rake the surface of the beds frequently to stop up these channels by which the sun is stealing the precious moisture.

The water that the surface soil cannot absorb sinks lower and lower into the ground. It finds no trouble to settle through layers of sand, for the particles do not fit closely together. It may come to a bed of clay which is far closer. Here progress is retarded. The water may accumulate, but finally it will get through, if the clay is not too closely packed. Again it may sink rapidly through thick beds of gravel or sand. Reaching another bed of clay which is stiffer by reason of the weight of the earth above it, the water may find that it cannot soak through. The only way to pass this clay barrier is to fill the basin, and to trickle over the edge, unless a place is found in the bottom where some looser substance offers a passage. Let us suppose that a concave clay basin of considerable depth is filled with water-soaked sand. At the very lowest point on the edge of this basin a stream will slowly trickle out, and will continue to flow, as long as water from above keeps the bowl full.

It is not uncommon to find on hillsides, in many regions, little brooks whose beginnings are traceable to springs that gush out of the ground. The spring fills a little basin, the overflow of which is the brook. If the source of this spring could be traced underground, we might easily follow it along some loose rock formation until we come to a clay basin like the one described above. We might have to go down quite a distance and then up again to reach the level of this supply, but the level of the water at the mouth of the spring can never be higher than the level of the water in the underground supply basin.

Often in hot summers springs "go dry." The level of water in the supply basin has fallen below the level of the spring. We must wait until rainfall has added to the depth of water in the basin before we can expect any flow into the pool which marks the place where the brook begins.

Suppose we had no beds of clay, but only sand and gravel under the surface soil. We should then expect the water to sink through this loose material without hindrance, and, finding its way out of the ground, to flow directly into the various branches of the main river system of our region. After a long rain we should have the streams flooded for a few days, then dry weather and the streams all low, many of them entirely dry until the next rainstorm.

Instead of this, the soil to a great depth is stored with water which cannot get away, except by the slow process by which the springs draw it off. This explains the steady flow of rivers. What should we do for wells if it were not for the water basins that lie below the surface? A shallow well may go dry. Its owner digs deeper, and strikes a lower "vein" of water that gives a more generous supply. In the regions of the country where the drift soil, left by the great ice-sheet, lies deepest, the glacial boulder clay is very far down. The surface water, settling from one level to another, finally reaches the bottom of the drift. Wells have to be deep that reach this water bed.

The water follows the slope of this bed and is drained into the ocean, sometimes by subterranean channels, because the bed of the nearest river is on a much higher level. So we must not think that the springs contain only the water that feeds the rivers. They contain more.

The layers of clay at different levels, from the surface down to the bottom of the drift, form water basins and make it possible for people to obtain a water supply without the expense of digging deep wells. The clayey subsoil, only a few feet below the surface, checks the downward course of the water, so that the sun can gradually draw it back, and keep a supply where plant roots can get it. The vapour rising keeps the air humid, and furnishes the dew that keeps all plant life comfortable and happy even through the hot summer months.

Under the drift lie layers of stratified rock, and under these are the granites and other fire-formed rocks, the beginning of those rock masses which form the solid bulk of the globe. We know little about the core of the earth, but the granites that are exposed in mountain ridges are found to have a great capacity for absorbing water, so it is not unlikely that much surface water soaks into the rock foundations and is never drained away into the sea.

The water in our wells is often hard. It becomes so by passing through strata of soil and rock made, in part, at least, of limestone, which is readily dissolved by water which contains some acid. Soil water absorbs acids from the decaying vegetation,—the dead leaves and roots of plants. Rain water is soft, and so is the water in ponds that have muddy basins, destitute of lime. Water in the springs and wells of the Mid-Western States is "hard" because it percolates through limestone material. In many parts of this country the well water is "soft," because of the scarcity of limestone in the soil.

I have seen springs around which the plants and the pebbles were coated with an incrustation of lime. "Petrified moss" is the name given to the plants thus turned to stone. The reason for this deposit is clear. Underground water is often subjected to great pressure, and at this time it is able to dissolve much more of any mineral substance than under ordinary conditions. When the pressure is released, the water is unable to hold in solution the quantity of mineral it contains; therefore, as it flows out through the mouth of the spring, the burden of mineral is laid down. The plants coated with the lime gradually decay, but their forms are preserved.

There are springs the water of which comes out burdened with iron, which is deposited as a yellowish or red mineral on objects over which it flows. Ponds fed by these springs accumulate deposits of the mineral in the muddy bottoms. Some of the most valuable deposits of iron ore have accumulated in bogs fed by iron-impregnated spring water. In a similar way lime deposits called marl or chalk are made.

THE SOIL IN FIELDS AND GARDENS

City and country teachers are expected to teach classes about the formation and cultivation of soil. It is surprising how much of the needed materials can be brought in by the children, even in the cities. The beginning is a flowering plant growing in a pot. A window box is a small garden. A garden plot is a miniature farm.

Materials to collect for study indoors. A few pieces of different kinds of rock: Granite, sandstone, slate; gravelly fragments of each, and finer sand. Pebbles from brooks and seashore. Samples of clays of different colors, and sands. Samples of sandy and clay soils, black pond muck, peat and coal. Rock fossils. A box of moist earth with earthworms in it. *Keep it moist.* A piece of sod, and a red clover plant with the soil clinging to its roots.

What is soil? It is the surface layer of the earth's crust, sometimes too shallow on the rocks to plough, sometimes much deeper. Under deep soil lies the "subsoil," usually hard and rarely ploughed.

What is soil made of? Ground rock materials and decayed remains of animal and plant life. By slow decay the soil becomes rich food for the growing of new plants. Wild land grows up to weeds and finally to forests. The soil in fields and gardens is cultivated to make it fertile. Plants take fertility from the soil. To maintain the same richness, plant food must be put back into the soil. This is done by deep tillage, and by mixing in with the soil manures, green crops, like clover, and commercial fertilizers.

Plants must be made comfortable, and must be fed. Few plants are comfortable in sand. It gets hot, it lets water through, and it shifts in wind and is a poor anchor for roots. Clay is so stiff that water cannot easily permeate it; roots have the same trouble to penetrate it and get at the food it is rich in. Air cannot get in.

Sand mixed with clay makes a mellow soil, which lets water and air pass freely through. The roots are more comfortable, and the tiny root hairs can reach the particles of both kinds of mineral food. But the needful third element is decaying plant and animal substances, called "humus." These enrich the soil, but they do a more important thing: their decay hastens the release of plant food from the earthy part of the soil, and they add to it a sticky element which has a wonderful power to attract and hold the water that soaks into the earth.

What is the best garden soil? A mixture of sand, clay, and humus is called "loam." If sand predominates, it is a sandy loam—warm, mellow soil. If clay predominates, we have a clay loam—a heavy, rich, but cool soil. All gradations between the two extremes are suited to the needs of crops, from the melons on sandy soil, to celery that prefers deep, cool soil, and cranberries that demand muck—just old humus.

How do plant roots feed in soil? By means of delicate root hairs which come into contact with particles of soil around which a film of soil water clings. This fluid dissolves the food, and the root absorbs the fluid. Plants can take no food in solid form. Hence it is of the greatest importance to have the soil pulverized and spongy, able to absorb and hold the greatest amount of water. The moisture-coated soil particles must have air-spaces between them. Air is as necessary to the roots as to the tops of growing plants.

Why does the farmer plough and harrow and roll the land? To pulverize the soil; to mellow and lighten it; to mix in thoroughly the manure he has spread on it, and to reach, if he can, the deeper layers that have plant food which the roots of his crops have not yet touched. Killing weeds is but a minor business, compared with tillage.

Later, ploughing or cultivating the surface lightly not only destroys the weeds, but it checks the loss of water by evaporation from the cracks that form in dry weather. Raking the garden once a day in dry weather does more good than watering it. The "dust mulch" acts as a cool sunguard over the roots.

The process of soil-making. If the man chopping wood in the Yosemite Valley looks about him he can see the soil-making forces at work on a grand scale. The bald, steep front of El Capitan is of the hardest granite, but it is slowly crumbling, and its fragments are accumulating at the bottom of the long slope. Rain and snow fill all crevices in the rocks. Frost is a wonderful force in widening these cracks, for water expands when it freezes. The loosened rock masses plough their way down the steep, gathering, as they go, increasing power to tear away any rocks in their path.

Wind blows finer rock fragments along, and they lodge in cracks. Fine dust and the seeds of plants are lodged there. The rocky slopes of the Yosemite Valley are all more or less covered with trees and shrubs that have come from wind-sown seeds. These plants thrust their roots deeper each year into the rock crevices. The feeding tips of roots secrete acids that eat away lime and other substances that occur in rocks. Dead leaves and other discarded portions of the trees rot about their roots, and form soil of increasing depth. The largest trees grow on the rocky soil deposited at the base of the slope. The tree's roots prevent the river from carrying it off.

When granite crumbles, its different mineral elements are separated. Clear, glassy particles of quartz we call sand. Dark particles of feldspar become clay, and may harden into slate. Sand may become sandstone. Exposed slate and sandstone are crumbled by exposure to wind and frost and moving water, and are deposited again as sand-bars and beds of clay.

The most interesting phase of soil study is the discovery of what a work the humble earthworm does in mellowing and enriching the soil.

THE WORK OF EARTHWORMS

The farmer and the gardener should expect very poor crops if they planted seed without first ploughing or spading the soil. Next, its fine particles must be separated by the breaking of the hard clods. A wise man ploughs heavy soil in the fall. It is caked into great clods which crumble before planting time. The water in the clods freezes in winter. The expansion due to freezing makes this soil water a force that separates the fine particles. So the frost works for the farmer.

Just under the surface of the soil lives a host of workers which are our patient friends. They work for their living, and are perhaps unconscious of the fact that they are constantly increasing the fertility of the soil. They are the earthworms, also called fishworms, which are distributed all over the world. They are not generally known to farmers and gardeners as friendly, useful creatures, and their services are rarely noticed. We see robins pulling them out of the ground, and we are likely to think the birds are ridding us of a garden pest. What we need is to use our eyes, and to read the wonderful discoveries recorded in a book called "Vegetable Mould and Earthworms," written by Charles Darwin.

The benefits of ploughing and spading are the loosening and pulverizing of the packed earth; the mixing of dead leaves and other vegetation on and near the surface with the more solid earth farther down; the letting in of water and air; and the checking of loss of water through cracks the sun forms by baking the soil dry.

The earthworm is a creature of the dark. It cannot see, but it is sufficiently sensitive to light to avoid the sun, the rays of which would shrivel up its moist skin. Having no lungs or gills, the worm uses the skin as the breathing organ; and it must be kept moist in order to serve its important use. This is why earthworms are never seen above ground except on rainy days, and never in the top soil if it has become dry. In seasons of little rain, they go down where the earth is moist, and venture to the surface only at night, when dew makes their coming up possible.

Earthworms have no teeth, but they have a long snout that protrudes beyond the mouth. Their food is found on and in the surface soil. They will eat scraps of meat by sucking the juices, and scrape off the pulp of leaves and root vegetables in much the same way. Much of their subsistence is upon organic matter that can be extracted from the soil. Quantities of earth are swallowed. It is rare that an earthworm is dug up that does not show earth pellets somewhere on their way through the long digestive canal. The rich juices of plant substance are absorbed from these pellets as they pass through the body.

Earthworms explore the surface of the soil by night, and pick up what they can find of fresh food. Nowhere have I heard of them as a nuisance in gardens, but they eagerly feed on bits of meat, especially fat, and on fresh leaves. They drag all such victuals into their burrows, and begin the digestion of the food by pouring on it from their mouths a secretion somewhat like pancreatic juice.

The worms honeycomb the earth with their burrows, which are long, winding tubes. In dry or cold weather these burrows may reach eight feet under ground. They run obliquely, as a rule, from the surface, and are lined with a layer of the smooth soil, like soft paste, cast from the body. The lining being spread, the burrow fits the worm's body closely. This enables it to pass quickly from one end to the other, though it must wriggle backward or forward without turning around.

At the lower end of the burrow, an enlarged chamber is found, where hibernating worms coil and sleep together in winter. At the top, a lining of dead leaves extends downward for a few inches, and in day time a plug of the same material is the outside door. At night the worm comes to the surface, and casts out the pellets of earth swallowed. The burrow grows in length by the amount of earth scraped off by the long snout and swallowed. The daily amount of excavation done is fairly estimated by the castings observed each morning on the surface.

One earthworm's work for the farmer is not very much, but consider how many are at work, and what each one is doing. It is boring holes through the solid earth, and letting in the surface water and the air. It is carrying the lower soil up to the surface, often the stubborn subsoil, that no plough could reach. It is burying and thus hastening the decay of plant fibre, which lightens heavy soil and makes it rich because it is porous. Moreover, the earthworms are doing over and over again this work of fining and turning over the soil, which the plough does but seldom.

By the continuous carrying up of their castings, the earthworms gradually bury manures spread on the surface. The collapse of their burrows and the making of new ones keep the soil constantly in motion. The particles are being loosened and brought into contact with the soil water, that dissolves, and thus frees for the use of feeding roots, the plant food stored in the rock particles that compose the mineral part of the soil.

The weight of earth brought to the surface by worms in the course of a year has been carefully estimated. Darwin gives seven to eighteen tons per acre as the lowest and highest reports, based on careful collecting of castings by four observers, working on small areas of totally different soils. In England, earthworms have done a great deal more toward burying boulders and ancient ruins than any other agency. They eagerly burrow under heavy objects, the weight of which causes them to crush the honeycombed earth. Undiscouraged, the earthworms repeat their work.

"Long before man existed, the land was regularly ploughed, and continues still to be ploughed by earthworms. It may be doubted whether there are many other animals which have played so important a part in the history of the world as have these lowly organized creatures."

After years of study, Charles Darwin came to this conclusion. The more we study the lives of these earth-consuming creatures, the more fully do we believe what the great nature student said. The fertile soil is made of rock meal and decayed leaves and roots. Only recently have ploughs been invented. But the great forest crops have grown in soil made mellow by the earthworm's ploughing.

QUIET FORCES THAT DESTROY ROCKS

Wind and water are the blustering active agents we see at work tearing down rocks and carrying away their particles. They do the most of this work of levelling the land; but there are quiet forces at work which might not attract our attention at all, and yet, without their help, wind and running water would not accomplish half the work for which they take the credit.

The air contains certain destructive gases which by their chemical action separate the particles of the hardest rocks, causing them to crumble. Now the wind blows away these crumbling particles, and the solid unchanged rock beneath is again exposed to the crumbling agencies.

The changes in temperature between day and night cause rocks to contract and expand, and these changes put a strain upon the mineral particles that compose them. Much scaling of rock surfaces is due to these causes. Building a fire on top of a rock, and then dashing water upon the heated mass, shatters it in many directions. This process merely intensifies the effect produced by the mild changes of winter and summer. Water is present in most rocks, in surprising quantities, often filling the spaces in porous rocks like sandstones.

When winter brings the temperature down to the freezing point, the water near the surface of the rock first feels it. Ice forms, and every particle of water is swollen by the change. A strain is put upon the mineral particles against which the particles of ice crowd for more room. Frost is a very powerful agent in the crumbling of rocks, as well as of stubborn clods of earth. In warm climates, and in desert regions where there is little moisture in the rocks, this destructive action of freezing water is not known. In cold countries, and in high altitudes, where the air is heavy with moisture, its greatest work is done.

Some kinds of rock decay when they become dry, and resist crumbling better when they absorb a certain amount of moisture. Alternate wetting and drying is destructive to certain rocks.

One of the unnoticed agents of rock decay is the action of lowly plants. Mosses grow upon the faces of rocks, thrusting their tiny root processes into pits they dig deeper by means of acids secreted by the delicate tips. You have seen shaded green patches of lichens, like little rugs, of different shapes, spread on the surface of rocks. But you cannot see so well the work these growths are doing in etching away the surface, and feeding upon the decaying mineral substance.

Mosses and lichens do a mighty work, with the help of water, in reducing rocks to their original elements, and thus forming soil. No plants but lichens and mosses can grow on the bare faces of rocks. As their root-like processes lengthen and go deeper into the rock face, particles are pried off, and the under-substance is attacked. Higher plants then find a footing. Have you not seen little trees growing on a patch of moss which gets its food from the air and the rock to which it clings? The spongy moss cushion soaks up the rain and holds it against the rock face. A streak of iron in the rock may cause the water to follow and rust it out, leaving a distinct crevice. Now the roots of any plant that happens to be growing on the moss may find a foot-hold in the crack. Streaks of lime in a rock readily absorb water, which gradually dissolves and absorbs its particles, inviting the roots to enter these new passages and feed upon the disintegrating minerals. Dead leaves decay, and the acids the trickling water absorbs from them are especially active in disintegrating lime rocks.

From such small beginnings has resulted the shattering of great rock masses by the growth of plants upon them. Tree roots that grow in rock crevices exert a power that is irresistible. The roots of smaller plants do the same great work in a quieter way.

When a hurricane or a flood tears down the mountain-side, sweeping everything before it, trees, torn out by the roots, drag great masses of rock and soil into the air, and fling them down the slope. Wind and water thus finish the destruction which the humble mosses and lichens began. What seemed an impregnable fortress of granite has crumbled into fragments. Its particles are reduced to dust, or are on the way to this condition. The plant food locked up in granite boulders becomes available to hungry roots. Forests, grain-fields, and meadows cover the work of destructive agencies with a mantle of green.

HOW ROCKS ARE MADE

The granite shaft is made out of the original substance of the earth's crust. Its minerals are the elements out of which all of the rock masses of the earth are formed, no matter how different they look from granite. Sandstone is made of particles of quartz. Clay and slate are made out of feldspar and mica. Iron ore comes from the hornblende in granite. The mineral particles, reassembled in different proportions, form all of the different rocks that are known.

Here in my hand is a piece of pudding-stone. It is made of pebbles of different sizes, each made of different coloured minerals. The pebbles are cemented together with a paste that has hardened into stone. This kind of rock the geologists call *conglomerate*. Pudding-stone is the common name, for the pebbles in the pasty matrix certainly do suggest the currants and the raisins that are sprinkled through a Christmas pudding.

Under the seashores there are forming to-day thick beds of sand. The rivers bring the rock material down from the hills, and it is sorted and laid down. The moving water drops the heaviest particles near shore, and carries the finer ones farther out before letting them fall.

The town of Cripple Creek, Colorado, which has grown up like magic since 1891, covers the richest gold and silver mines in the world

The level valley is filled up with fine rock flour washed from the sides of the neighboring mountains

The hard water, that comes through limestone rocks, adds lime in solution to the ocean water. All the shellfish of the sea, and the creatures with bony skeletons, take in the bone-building, shell-making lime with their food. Generations of these inhabitants of the sea have died, and their shells and bones have accumulated and been transformed into thick beds of limestone on the ocean floor. This is going on to-day; but the limestone does not accumulate as rapidly as when the ocean teemed with shell-bearing creatures of gigantic size. Of these we shall speak in another chapter.

The fine dust that is blown into the ocean from the land, and that makes river water muddy, accumulates on the sea bottom as banks of mud, which by the burden of later deposits is converted into clay. Sandstone is but the compressed sand-bank.

In the study of mountains, geologists have discovered that old seashores were thrown up into the first great ridges that form the backbone of a mountain system. The Rocky Mountains, and the Appalachian system on the east, were made out of thick strata of rocks that had been formed by accumulations of mud and sand—the washings of the land—on the opposite shores of a great mid-continental sea, that stretched from the crest of one great mountain system across to the other, and north and south from the Laurentian Hills to the Gulf of Mexico. The great weight of the accumulating layers of rock materials on one side, and the wasted land surfaces on the other, made the sea border a line of greatest weakness in the crust of the earth. The shrinking of the globe underneath caused the break; mashing and folding followed, throwing the ridge above sea-level, and making dry land out of rock waste which had been accumulating, perhaps for millions of years, under the sea. The wrinkling of the earth's crust was the result of crushing forces which produced tremendous heat.

Streams of lava sprang out through the fissures and poured streams of melted rock down the sides of the fold, quite burying, in many places, the layers of limestone, sandstone, and clay. Between the strata of water-formed rocks there were often created chimney-like openings, into which molten rock from below was forced, forming, when cool, veins and dikes of rock material, specimens of the substance of the earth's interior.

Tremendous pressure and heat, acting upon stratified rocks saturated with water transform them into very different kinds of rock. Limestone, subjected to these forces, is changed into marble. Clays are transformed into slates. Sandstone is changed into quartzite, the sand grains being melted so as to become no longer visible to the naked eye. The anthracite coal of the Pennsylvania mountains is the result of heat and pressure acting upon soft coal. Associated with these beds of hard coal are beds of black lead, or graphite, the substance used in making "lead" pencils. We believe that the same forces that operated to transform clay rocks into slate, and limestone into marble, transformed soft coal into hard, and hard coal into graphite, in the days when the earth was young.

The word *sedimentary* is applied to rocks which were originally laid down under water, as sediment, brought by running water, or by wind, or by the decay of organic substances. *Stratified* rocks are those which are arranged in layers. Sedimentary rocks will fall into this class. *Aqueous* rocks are those which are formed under water. Most of the stratified and sedimentary rocks, but not all, may be included under this term. Rocks that are made out of fragments of other rocks torn down by the agencies of erosion are called *fragmental*. Wind, water, and ice are the three great agencies that wear away the land, bring rock fragments long distances, and deposit them where aqueous rocks are being formed. Volcanic eruptions bring material from the earth's interior. This material ranges all the way from huge boulders to the finest impalpable dust, called volcanic ashes. Rivers of ice called glaciers crowd against their banks, loosening rock masses and carrying away fragments of all sizes, in their progress down the valley. Brooks and rivers carry the pebbles and the larger rock masses they are able to loosen from their walls and beds, and grind them smooth as they move along toward lower levels.

The air itself causes rocks to crumble; percolating water robs them of their soluble salts, reducing even solid granite to a loose mass of quartz grains and clay. Plants and animals absorb as food the mineral substances of rocks, when they are dissolved in water. They transform these food elements into their own body substance, and finally give back their dead bodies, the mineral substances of which are freed by decay to return to the earth, and become elements of rock again.

The decay of rock is well shown by the materials that accumulate at the base of a cliff. Angular fragments of all sizes, but all more or less flattened, come from strata of shaly rock, that can be seen jutting out far above. A great deal of this sort of material is found mingled with the soil of the Northeastern States. Round pebbles in pudding-stone have been formed in brook beds and deposited on beaches where they have become caked in mud and finally consolidated into rock. If the beach chanced to be sandy instead of muddy, a matrix of sandy paste holds the larger pebbles in place. Limestone paste cements together the pebbles of limestone conglomerates.

In St. Augustine many of the houses are built of coquina rock, a mass of broken shells which have become cemented together by lime mud, derived from their own decay. On the slopes of volcanoes, rock fragments of all kinds are cemented together by the flowing lava. So we see that there are pudding-stones of many kinds to be found. If some solvent acid is present in the water that percolates through these rocks it may soften the cement and thus free the pebbles, reducing the conglomerate again to a mere heap of shell fragments, or gravel, or rounded pebbles.

The story of rock formation tells how fire and water, and the two combined, have made, and made over, again and again, the substance of the earth's crust. Chemical and physical changes constantly tear down some portions of the earth to build up others. The constant, combined effort of wind and water is to level the earth and fill up the ocean bed. Rocks are constantly being formed; the changes that have been going on since the world began are still in progress. We can see them all about us on any and every day of our lives.

GETTING ACQUAINTED WITH A RIVER

I have two friends whose childhood was spent in a home on the banks of a noble eastern river. Their father taught the boy and the girl to row a boat, and later each learned the more difficult art of managing a canoe. On holidays they enjoyed no pleasure so much as a picnic on the river-bank at some point that could be reached by rowing. As they grew older, longer trips were planned, and the river was explored as far as it was navigable by boat or canoe. Last summer when vacation came, these two carried out a long-cherished plan to find the beginning of the river—to follow it to its source. So they left home, and canoed up-stream, until the stream became a brook, so shallow they could go no farther. Then they followed it on foot—wading, climbing, making little détours, but never losing the little river. At last they came to the beginning of it—a tiny rivulet trickled out of the side of a hill, filling a wooden keg that formed a basin, where thirsty passers-by could stoop and drink. They decided to mark the spring, so that people who found it later, and were refreshed by its clear water, might know that here was born the greatest river of a great state. But they were not the original discoverers. Above the spring, a board was nailed to a tree, saying that this is the headwater of the river with the beautiful Indian name, Susquehanna.

It was a dry summer, and the overflow of the basin was almost all drunk up by the thirsty ground. They could scarcely follow it, except by the groove cut by the rivulet in seasons when the flow was greater. They followed the runaway brook, through the grass roots, that almost hid it. As the ground grew steeper, it hurried faster. Soon it gathered the water of other springs, which hurried toward it in small rivulets, because its level was lower. Water always seeks the lowest level it can find. Sometimes marshy spots were reached where water stood in the holes made by the feet of cattle that came there to drink. The water was muddy, and seemed to stand still. But it was settling steadily, and at one side the little river was found, flowing away with the water it drew from the swampy, springy ground. All the mud was gone, now; the water was clear. It flowed in a bed with a stony floor, and there were rough steps where the water fell down in little sheets, forming a waterfall, the first of many that make this river beautiful in the upper half of its course. To get from the high level of that hillside spring to the low level of the sea, the water has to make a fall of twenty-three hundred feet, but it makes the descent gradually. It could not climb over anything, but always found a way to get around the rocks and hills that stood in its way. When the flat marsh land interfered, the water poured in and overflowed the basin at the lowest margin.

In the rocky ground the two explorers found that the stream had widened its channel by entering a narrow crevice and wearing away its walls. The continual washing of the water wears away stone. Rocks are softened by being wet. Streaks of iron in the hardest granite will rust out and let the water in. Then the lime in rocks is easily dissolved. Every dead leaf the river carried along added an acid to the water, and this made easier the process of dissolving the limestone.

Every crumbling rock gives the river tools that it uses like hammer and chisel and sandpaper to smooth all the uneven surfaces in its bed, to move stumbling blocks, and to dig the bed deeper and wider. The steeper the slope is, the faster the stream flows, and the larger the rocks it can carry. Rocks loosened from the stream bed are rolled along by the current. Then bang! against the rocks that are not loose, and often they are able to break them loose. The fine sand is swept along, and its sharp points strike like steel needles, and do a great work in polishing roughness and loosening small particles from the stream bed. The bigger pebbles of the stream have banged against the rock walls, with the same effect, smoothing away unevenness and pounding fragments loose, rolling against one another, and getting their own rough corners worn away.

The makers of stone marbles learned their business from a brook. They cut the stone into cubical blocks, and throw them into troughs, into which is poured a stream of running water. The blocks are kept in motion, and the grinding makes each block help the rest to grind off the eight corners and the twelve ridges of each one. The water becomes muddy with the fine particles, just as the drip from a grindstone becomes unclean when an axe is ground. Pretty soon all the blocks in the trough are changed into globes—the marbles that children buy at the shops when marble season comes around.

I suppose if the troughs are not watched and emptied in time, the marbles would gradually be ground down to the size of peas, then to the size of small bird shot, and finally they would escape as muddy water and fine sand grains.

Sure it is that the sandy shores that line most rivers are the remnants of hard rocks that have been torn out and ground up by the action of the current.

Not very many miles from its first waterfall the stream had grown so large that my two friends knew that they would soon find their canoes. The plan now was to float down the curious, winding river and to learn, if the river and the banks could tell them, just why the course was so crooked on the map. They came into a broad, level valley where streams met them, coming out of deep clefts between the hills they were leaving behind them.

The banks were pebbly, but blackened with slimy mud that made the water murky. The current swerved from one side to the other, sometimes quite close to the bank, where the river turned and formed a deep bend. On this side the bank was steep, the roots of plants and trees exposed. On the opposite side a muddy bank sloped gently out into the stream. Here building up was going on, to offset the tearing down.

The sharp bends are made sharper, once the current is deflected from the middle of the stream to one side. At length the loops bend on each other and come so near together that the current breaks through, leaving a semicircular bayou of still water, and the river's course straightened at that place. It must have been in a spring flood that this cut-off was made, and, the break once made was easily widened, for the soil is fine mud which, when soaked, crumbles and dissolves into muddy water.

Stately and slow that river moves down to the bay, into which it empties its load. The rain that falls on hundreds of square miles of territory flows into the streams that feed this trunk. The little spring that is the headwater of the system is but one of many pockets in the hillsides that hold the water that soaks into the ground and give it out by slow degrees. Surface water after a rain flows quickly into the streams. It is the springs that hold back their supply and keep the rivers from running dry in hot weather.

Do they feel now that they know their river? Are they ready to leave it, and explore some other? Indeed, no. They are barely introduced to it. All kinds of rivers are shown by the different parts of this one. It is a river of the mountains and of the lowland. It flows through woods and prairies, through rocky passes and reedy flats. It races impetuously in its youth, and plods sedately in later life. The trees and the other plants that shadow this stream, and live by its bounty, are very different in the upland and in the lowland. The scenery along this stream shows endless variety. Up yonder all is wild. Down here great bridges span the flood, boats of all kinds carry on the commerce between two neighbour cities. A great park comes down to the river-bank on one side. Canoes are thick as they can paddle on late summer afternoons.

No one can ever really know a river well enough to feel that it is an old story. There is always something new it has to tell its friends. So my two explorers say, and they know far more about their friendly river than I do.

THE WAYS OF RIVERS

A canal is an artificial river, built to carry boats from one place to another. Its course is, as nearly as possible, a straight line between two points. A river, we all agree, is more beautiful than a canal, for it winds in graceful curves, in and out among the hills, its waters seeking the lowest level, always.

No artist could lay out curves more beautiful than the river forms. These curves change from year to year, some slowly, some more rapidly. It is not hard to understand just why these changes take place.

Some rivers are dangerous for boating at certain points. The current is strong, and there are eddies and whirlpools that have to be avoided, or the boat becomes unmanageable. People are drowned each season by trusting themselves to rivers the dangerous tricks of which they do not know. Deep holes are washed out of the bed of the stream by whirling eddies. The pot-holes of which people talk are deep, rounded cavities, ground out of the rocky stream-bed by the scouring of sand and loose stones driven by whirling eddies in shallow basins. Every year deepens each pot-hole until some change in the stream-bed shifts the eddy to another place.

No stream finds its channel ready-made; it makes its own, and constantly changes it. The current swings to one side of the channel, lifting the loose sediment and grinding deeper the bed of the stream. The water lags on the opposite side, and sediment falls to the bottom. So the building-up of one side is going on at the same time that the tearing-down process is being carried on on the other. With the lowering of the bed the river swerves toward one bank, and a hollow is worn by slow degrees. The current swings into this hollow, and in passing out is thrown across the stream to the opposite bank. Here its force wears away another hollow; and so it zigzags down-stream. The deeper the hollows, the more curved becomes the course, if the general fall is but moderate. It is toward the lower courses of the stream that the winding becomes more noticeable. The sediment that is carried is deposited at the point where the current is least strong, so that while the outcurves become sharper by the tearing away of the stream's bank, the incurves become sharper by the building up of this bank

The Mississippi below Memphis is thrown into a wonderful series of curves by the erosion and the deposit caused by the current zigzagging back and forth from one bank to the other. Gradually the curves become loops. The river's current finally jumps across the meeting of the curves, and abandons the circular bend. It becomes a bayou or lagoon of still water, while the current flows on in the straightened channel. All rivers that flow through flat, swampy land show these intricate winding channels and many lagoons that have once been curves of the river.

No one would ever mistake a river for a lake or any other body of water, yet rivers differ greatly in character. One tears its way along down its steep, rock-encumbered channel between walls that rise as vertical precipices on both sides. The roaming, angry waters are drawn into whirlpools in one place. They lie stagnant as if sulking in another, then leap boisterously over ledges of rock and are churned into creamy foam at the bottom. Outside the mountainous part of its course this same river flows broad and calm through a mud-banked channel, cut by tributary streams that draw in the water of low, sloping hills.

The Missouri is such a wild mountain stream at its headwaters. We who have seen its muddy waters from Sioux City to St. Louis would hardly believe that its impetuous and picturesque youth could merge into an old age so comfortable and placid and commonplace.

This thing is true of all rivers. They flow, gradually or suddenly, from higher to lower levels. To reach the lowest level as soon as possible is the end each river is striving toward. If it could, each river would cut its bed to this depth at the first stage of its course. Its tools are the rocks it carries, great and small. The force that uses these tools is the power of falling water, represented by the current of the stream. The upper part of a river such as the Missouri or Mississippi engages in a campaign of widening and deepening its channel, and carrying away quantities of sediment. The lower reaches of the stream flow through more level country; the current is checked, and a vast burden of sediment is laid down. Instead of tearing away its banks and bottom, the river fills up gradually with mud. The current meanders between banks of sediment over a bottom which becomes shallower year by year. The Rocky Mountains are being carried to the Gulf of Mexico. The commerce of the river is impeded by mountain débris deposited as mud-banks along the river's lower course.

Many rivers are quiet and commonplace throughout their length. They flow between low, rounded hills, and are joined by quiet streams, that occupy the separating grooves between the hills. This is the oldest type of river. It has done its work. Rainfall and stream-flow have brought the level of the land nearly to the level of the stream. Very little more is left to be ground down and carried away. The landscape is beautiful, but it is no longer picturesque. Wind and water have smoothed away unevennesses. Trees and grass and other vegetation check erosion, and the river has little to do but to carry away the surface water that falls as rain.

But suppose our river, flowing gently between its grassy banks, should feel some mighty power lifting it up, with all its neighbour hills and valleys, to form a wrinkle in the still unstable crust of the earth. Away off at the river's mouth the level may not have changed, or that region may have been depressed instead of elevated by the shrinking process. Suppose the great upheaval has not severed the upper from the lower courses of the stream. With tremendous force and speed, the current flows from the higher levels to the lower. The river in the highlands strikes hard to reach the level of its mouth. It grinds with all its might, and all its rocky tools, upon its bed. All the mud is scoured out, and then the underlying rocks are attacked. If these rocks are soft and easily worn away, the channel deepens rapidly. One after another the alternating layers are excavated, and the river flows in a canyon which deepens more and more. As the level is lowered, the current of the stream becomes slower and the cutting away of its bed less rapid. The stream is content to flow gently, for it has almost reached the old level, on which it flowed before the valley became a ridge or table-land.

The rivers that flow in canyons have been thousands of years in carving out their channels, yet they are newer, geologically speaking, than the streams that drain the level prairie country. The earth has risen, and the canyons have been carved since the prairies became rolling, level ground.

This little pond is a basin hollowed by the same glacier that scattered the stones and rounded the hills

Every stream is wearing away its banks, while trees and grass blades are holding on to the soil with all their roots

The Colorado River flows through a canyon with walls that in places present sheer vertical faces a mile in depth, and so smooth that no trail can be found by which to reach from top to bottom. The region has but slight erosion by wind, and practically none by rain. The local rainfall is very slight. So the river is the one force that has acted to cut down the rocks, and its force is all expended in the narrow area of its own bed. Had frequent rains been the rule on the Colorado plateau, the angles of the mesas would have been rounded into hills of the familiar kind so constantly a part of the landscape in the eastern half of the continent.

The Colorado is an ancient river which has to carry away the store of moisture that comes from the Pacific Ocean and falls as snow on the high peaks of the Rocky Mountains. Similar river gorges with similar stories to tell are the Arkansas, the Platte, and the Yellowstone. All cut their channels unaided through regions of little rain.

When the earth's crust is thrown up in mountain folds, and between them valleys are formed, the level of rivers is sometimes lowered and the rapidity of their flow is checked. A stream which has torn down its walls at a rapid rate becomes a sluggish water-course, its current clogged with sediment, which it has no power to carry farther. When such a river begins to build and obstruct its own waters it bars its progress and may form a lake as the outlet of its tributary streams. Many ancient rivers have been utterly changed and some obliterated by general movements of the earth's crust.

THE STORY OF A POND

Look out of the car window as you cross a flat stretch of new prairie country, and you see a great many little ponds of water dotting the green landscape. Forty years ago Iowa was a good place to see ponds of all shapes and sizes. The copious rainfall of the early spring gathered in the hollows of the land, and the stiff clay subsoil prevented the water from soaking quickly into the ground. The ponds might dry away during the hot, dry summer, leaving a baked clay basin, checked with an intricate system of cracks. Or if rains were frequent and heavy, they might keep full to the brim throughout the season.

Tall bulrushes stood around the margins of the largest ponds, and water-lilies blossomed on the surface during the summer. The bass and the treble of the spring chorus were made by frogs and toads and little hylas, all of which resorted to the ponds to lay their eggs, in coiled ropes or spongy masses, according to their various family traditions. On many a spring night my zoölogy class and I have visited the squashy margins of these ponds, and, by the light of a lantern, seen singing toads and frogs sitting on bare hummocks of grass roots that stood above the water-line. The throat of each musician was puffed out into a bag about the size and shape of a small hen's egg; and all were singing for dear life, and making a din that was almost ear-splitting at close range. So great was the self-absorption of these singers that we could approach them, daze them with the light of the lantern, and capture any number of them with our long-handled nets before they noticed us. But it was not easy to persuade them to sing in captivity, no matter how many of the comforts of home we provided in the school aquariums. So, after some very interesting nature studies, we always carried them back and liberated them, where they could rejoin their kinsfolk and neighbours.

It was when we were scraping the mud from our rubber boots that we realized the character of the bottoms of our prairie ponds. The slimy black deposit was made partly of the clay bottom, but largely of decaying roots and tops of water plants of various kinds. Whenever it rained or the wind blew hard, the bottom was stirred enough to make the water muddy; and on the quietest days a pail of pond water had a tinge of brown because there were always decaying leaves and other rubbish to stain its purity.

The farmers drained the ponds as fast as they were able, carrying the water, by open ditches first, and later by underground tile drains, to lower levels. Finally these trunk drain pipes discharged the water into streams or lakes. To-day a large proportion of the pond areas of Iowa has disappeared; the hollow tile of terra-cotta has been the most efficient means of converting the waste land, covered by ponds, into fertile fields.

But the ponds that have not been drained are smaller than they used to be, and are on the straight road to extinction. This process one can see at any time by visiting a pond. Every year a crop of reeds and a dozen other species of vigorous water plants dies at the top and adds the substance of their summer growth to the dust and other refuse that gathers in the bottom of the pond. Each spring roots and seeds send up another crop, if possible more vigorous than the last, and this top growth in turn dies and lies upon the bottom. The pond level varies with the rainfall of the years, but it averages a certain depth, from which something is each year subtracted by the accumulations of rotting vegetable matter in the bottom. Evaporation lowers the water-level, especially in hot, dry summers. From year to year the water plants draw in to form a smaller circle, the grassy meadow land encroaches on all sides. The end of the story is the filling up of the pond basin with the rotting substance of its own vegetation. This is what is happening to ponds and inland marshes by slow degrees. The tile drain pipes obliterate the pond in a single season. Nature is more deliberate. She may require a hundred years to fill up a single pond which the farmer can rid himself of by a few days of work and a few rods of tiling.

THE RIDDLE OF THE LOST ROCKS

Outside of my window two robins are building a nest in the crotch of a blossoming red maple tree. And just across the hedge, men are digging a big square hole in the ground—the cellar of our neighbour's new house. It looks now as if the robins would get their house built first, for they need but one room, and they do not trouble about a cellar. I shall watch both houses as they grow through the breezy March days.

The brown sod was first torn up by a plough, which uncovered the red New Jersey soil. Two men, with a team hitched to a scraper, have carried load after load of the loose earth to a heap on the back of the lot, while two other men with pickaxes dug into the hard subsoil, loosening it, so that the scraper could scoop it up.

This subsoil is heavy, like clay, and it breaks apart into hard clods. At the surface the men found a network of tree roots, about which the soil easily crumbled. Often I hear a sharp, metallic stroke, unlike the dull sound of the picks striking into the earth. The digger has struck a stone, and he must work around it, pry it up and lift it out of the way. A row of these stones is seen at one side of the cellar hole, ranged along the bank. They are all different in size and shape, and red with clay, so I can't tell what they are made of. But from this distance I see plainly that they are irregular in form and have no sharp corners. The soil strewn along the lot by the scraper is full of stones, mostly irregular, but some rounded; some are as big as your head, others grade down to the sizes of marbles.

When I went down and examined this red earth, I found pebbles of all shapes and sizes, gravel in with the clay, and grains of sand. This rock-sprinkled soil in New Jersey is very much like soil which I know very well in Iowa; it looks different in colour, but those pebbles and rock fragments must be explained in the same way here as there.

These are not native stones, the outcrop of near-by hillsides, but strangers in this region. The stones in Iowa soil are also imported.

The prairie land of Iowa has not many big rocks on the surface, yet enough of them to make trouble. The man who was ploughing kept a sharp lookout, and swung his plough point away from a buried rock that showed above ground, lest it should break the steel blade. One of the farmer's jobs for the less busy season was to go out with sledge and dynamite sticks, and blast into fragments the buried boulders too large to move. Sometimes building a hot fire on the top of it, and throwing on water, would crack the stubborn "dornick" into pieces small enough to be loaded on stone-boats.

I remember when the last giant boulder whose buried bulk scarcely showed at the surface, was fractured by dynamite. Its total weight proved to be many tons. We hauled the pieces to the great stone pile which furnished materials for walling the sides of a deep well and for laying the foundation of the new house. Yet for years stones have been accumulating, all of them turned out of the same farm, when pastures and swampy land came under the plough.

Draw a line on the map from New York to St. Louis, and then turn northward a little and extend it to the Yellowstone Park. The boulder-strewn states lie north of this line, and are not found south of it, anywhere. Canada has boulders just like those of our Northern States. The same power scattered them over all of the vast northern half of North America and a large part of Europe.

What explanation is there for this extensive distribution of unsorted débris?

THE QUESTION ANSWERED

The rocks tell their own story, partly, but not wholly. They told just enough to keep the early geologists guessing; and only very recently has the guessing come upon the truth.

These things the rocks told:

1. We have come from a distance.

2. We have had our sharp corners worn off.

3. Many of us have deep scratches on our sides.

4. At various places we have been dumped in long ridges, mixed with much earth.

5. A big boulder is often balanced on another one.

The first thing the geologist noted was the fact that these boulders are strangers—that is, they are not the native rocks that outcrop on hillsides and on mountain slopes near where they are found. Far to the north are beds of rock from which this débris undoubtedly came. Could a flood have scattered them as they are found? No, for water sorts the rock débris it deposits, and it rounds and polishes rock fragments, instead of scratching and grooving them and leaving them angular, as these are.

Professor Agassiz went to Switzerland and studied the glaciers. He found unsorted rock fragments where the glacier's nose melted, and let them fall. They were worn and scratched and grooved, by being frozen into the ice, and dragged over the rocky bed of the stream. The rocky walls of the valley were scored by the glacier's tools. Rounded domes of rock jutted out of the ground, in the paths of the ice streams, just like the granite outcrop in Central Park in New York, and many others in the region of scattered boulders.

After long studies in Europe and in North America, Professor Agassiz declared his belief that a great ice-sheet once covered the northern half of both countries, rounding the hills, scooping out the valleys and lake basins, and scattering the boulders, gravel, and clay, as it gradually melted away.

The belief of Professor Agassiz was not accepted at once, but further studies prove that he guessed the riddle of the boulders. The rich soil of the Northern States is the glacial drift—the mixture of rock fragments of all sizes with fine boulder clay, left by the gradual melting of the great ice-sheet as it retreated northward at the end of the "Glacial Epoch."

GLACIERS AMONG THE ALPS

Switzerland is a little country without any seacoast, mountainous, with steep, lofty peaks, and narrow valleys. The climate is cool and moist, and snow falls the year round on the mountain slopes. A snow-cap covers the lower peaks and ridges. Above the level of nine thousand feet the bare peaks rise into a dry atmosphere; but below this altitude, and above the six thousand-foot mark, lies the belt of greatest snowfall. Peaks between six and nine thousand feet high are buried under the Alpine snow-field, which adds thickness with each storm, and is drained away to feed the rushing mountain streams in the lower valleys.

The snow that falls on the steep, smooth slope clings at first; but as the thickness and the weight of these snow banks increase, their hold on the slope weakens. They may slip off, at any moment. The village at the foot of the slope is in danger of being buried under a snow-slide, which people call an avalanche. "Challanche" is another name for it. The hunter on the snow-clad mountains dares not shout for fear that his voice, reëchoing among the silent mountains, may start an avalanche on its deadly plunge into the valley.

On the surface of the snow-field, light snow-flakes rest. Under them the snow is packed closer. Deeper down, the snow is granular, like pellets of ice; and still under this is ice, made of snow under pressure. The weight of the accumulated snow presses the underlying ice out into the valleys. These streams are the glaciers—rivers of ice.

The glaciers of the Alps vary in length from five to fifteen miles, from one to three miles in width, and from two hundred to six hundred feet in thickness. They flow at the rate of from one to three feet a day, going faster on the steeper slopes.

It is hard to believe that any substance as solid and brittle as ice can flow. Its movement is like that of stiff molasses, or wax, or pitch. The tremendous pressure of the snow-field pushes the mass of ice out into the valleys, and its own weight, combined with the constant pressure from behind, keeps it moving.

The glacier's progress is hindered by the uneven walls and bed of the valley, and by any decrease in the slope of the bed. When a flat, broad area is reached, a lake of ice may be formed. These are not frequent in the Alps. The water near the banks and at the bottom of a river does not flow as swiftly as in the middle and at the surface of the stream. The flow of ice in a glacier is just so. Friction with the banks and bottom retards the ice while the middle parts go forward, melting under the strain, and freezing again. There is a constant readjusting of particles, which does not affect the solidity of the mass.

The ice moulds itself over any unevenness in its bed if it cannot remove the obstruction. The drop which would cause a small waterfall in a river, makes a bend in the thick body of the ice river. Great cracks, called *crevasses*, are made at the surface, along the line of the bend. The width of the V-shaped openings depends upon the depth of the glacier and the sharpness of the bend that causes the breaks.

Rocky ridges in the bed of the ice-stream may cause crevasses that run lengthwise of the glacier. Snow may fill these chasms or bridge them over. The hunter or the tourist who ventures on the glacier is in constant danger, unless he sees solid ice under him. Men rope themselves together in climbing over perilous places, so that if one slips into a crevasse his mates can save him.

A glacier tears away and carries away quantities of rock and earth that form the walls of its bed. As the valley narrows, tremendous pressure crowds the ice against the sides, tearing trees out by the roots and causing rock masses to fall on the top of the glacier, or to be dragged along frozen solidly into its sides. The weight of the ice bears on the bed of the glacier, and its progress crowds irresistibly against all loose rock material. The glacier's tools are the rocks it carries frozen into its icy walls and bottom. These rocks rub against the walls, grinding off débris which is pushed or carried along. No matter how heavy the boulders are that fall in the way of the ice river, the ice carries them along. It cannot drop them as a river of water would do. Slowly they travel, and finally stop where the nose of the glacier melts and leaves all débris that the mountain stream, fed by the melting of the ice, cannot carry away.

The bedrock under a glacier is scraped and ground and scored by the glacier's tools the rock fragments frozen into the bottom of the ice. These rocks are worn away by constant grinding, just as a steel knife becomes thin and narrow by use. Scratches and scorings and polished surfaces are found in all rocks that pass one another in close contact. Its worn-out tools the glacier drops at the point where its ice melts. This great, unsorted mass of rock meal and coarser débris the stream is gradually scattering down the valley.

The name "moraine" has been given to the earth rubbish a glacier collects and finally dumps. The *top moraine* is at the surface of the ice. The *lateral moraines*, one at each side, are the débris gathered from the sides of the valley. The *ground moraine* is what débris the ice pushes and drags along on the bottom. The *terminal moraine* is the dumping-ground of this mass of material, where the ice river melts.

Glaciers, like other rivers, often have tributary streams. A *median moraine*, seen as a dark streak running lengthwise on the surface of a glacier, means that two branch glaciers have united to form this one. Go back far enough and you will reach the place where the two streams come together. The two lateral moraines that join form the middle line of débris, the median moraine. Three ice-streams joined produce two top moraines. They locate the lateral moraines of the middle glacier.

The surface of a glacier is often a mass of broken and rough ice, forming a series of pits and pinnacles that make crossing impossible. The sun melts the surface, forming pools and percolating streams of water, that honeycomb the mass. Underneath, the ice is tunnelled, and a rushing stream flows out under the end of the glacier. It is not clear, but black with mud, called *boulder clay*, or *till*, made of ground rock, and mixed with fragments of all shapes and sizes. This is the meal from the glacier's mill, dumped where the water can sift it.

"Balanced rocks" are boulders, one upon another, that once lay on a glacier, and were left in this strange, unstable position when the supporting ice walls melted away from them. In Bronx Park in New York the "rocking stone" always attracts attention. The glacier that lodged it there, also rounded the granite dome in Central Park and scattered the rock-strewn boulder clay on Long Island. Doubtless in an earlier day the edges of this glacier were thrust out into the Atlantic, not far from the Great South Bay, and icebergs broke off and floated away.

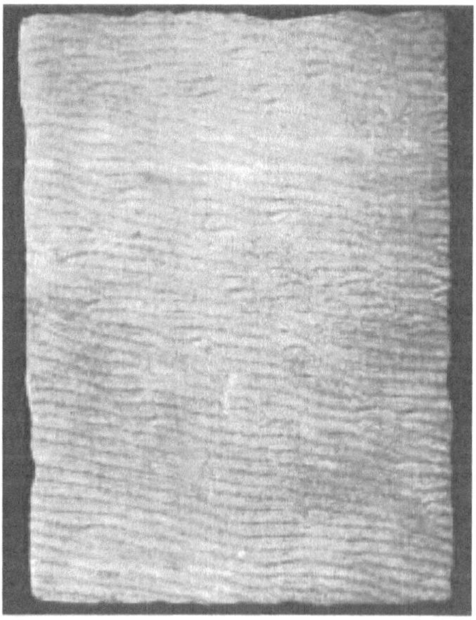

Potsdam sandstone showing ripple marks

By permission of the American Museum of Natural History
Glacial striæ on Lower Helderberg limestone

Glacial grooves in the South Meadow, Central Park, New York

By permission of the American Museum of Natural History
Mt. Tom, West 83d St., New York

Glaciers are small to-day compared with what they were long ago, in Europe and in America. The climate became warmer, and the ice-cap retreated. Old moraines show that the ice rivers of the Alps once came much farther down the valleys than they do now. Smooth, deeply scored domes of rock, the one in Central Park and the bald head of Mount Tom, are just like those that lie in Alpine valleys from which the glaciers have long ago retreated. There are old moraines far up the sides of valleys, showing that once the glaciers were far deeper than now. No other power could have brought rocks from strata higher up the mountains, and lodged them thus.

Nearer home, Mt. Shasta and Mt. Rainier still have glaciers that have dwindled in size, until they bear little comparison to the gigantic ice-streams that once filled the smooth beds their puny successors flow into. Remnants of glaciers lie in the hollows of the Sierras. We must go north to find the snow-fields of Alaska and glaciers worthy to be compared with those ancient ice rivers whose work is plainly to be seen, though they are gone.

THE GREAT ICE-SHEET

Greenland is green only along its southern edge, and only in summer, so its name is misleading. It is a frozen continent lying under a great ice-cap, which covers 500,000 square miles and is several thousand feet in thickness. The top of this icy table-land rises from five thousand to ten thousand feet above the sea-level. The long, cold winters are marked by great snowfall, and the drifts do not have time to melt during the short summer; and so they keep getting deeper and deeper. Streams of ice flow down the steeps into the sea, and break off by their weight when they are pushed out into the water. These are the icebergs which float off into the North Atlantic, and are often seen by passengers on transatlantic steamers.

Long ago Greenland better deserved its name. Explorers who have climbed the mountain steeps that guard the unknown ice-fields of the interior have discovered, a thousand feet above the sea-level, an ancient beach, strewn with shells of molluscs like those which now inhabit salt water, and skeletons of fishes lie buried in the sand. It is impossible to think that the ocean has subsided. The only explanation that accounts for the ancient beach, high and dry on the side of Greenland's icy mountain is that the continent has been lifted a thousand feet above its former level. This is an accepted fact.

We know that climate changes with changed altitude as well as latitude. Going up the side of a mountain, even in tropical regions, we may reach the snow-line in the middle of summer. Magnolia trees and tree ferns once grew luxuriantly in Greenland forests. Their fossil remains have been found in the rocks. This was long before the continent was lifted into the altitude of ice and snow. And it is believed that the climate of northern latitudes has become more severe than formerly from other causes. It is possible that the earth's orbit has gradually changed in form and position.

If Greenland should ever subside until the ancient beach rests again at sea-level, the secrets of that unknown land would be revealed by the melting of the glacial sheet that overspreads it. Possibly it would turn out to be a mere flock of islands. We can only guess. North America had, not so long ago, two-thirds of its area covered with an ice-sheet like that of Greenland, and a climate as cold as Greenland's. At this time the land was lifted two to three thousand feet higher than its present level. All of the rain fell as snow, and the ice accumulated and became thicker year by year. Instead of glaciers filling the gorges, a great ice flood covered all the land, and pushed southward as far as the Ohio River on the east and Yellowstone Park in the west. The Rocky Mountains and some parts of the Appalachian system accumulated snow and formed local glaciers, separated from the vast ice-sheet.

The unstable crust of the earth began to sink at length, and gradually the ice-sheet's progress southward was checked, and it began to recede by melting. All along the borders of this great fan-shaped ice-field water accumulated from the melting, and flooded the streams which drained it to the Atlantic and the Gulf. Icebergs broken off of the edge of the retiring ice-sheet floated in a great inland sea. The land sank lower and lower until the general level was five hundred to one thousand feet lower than it now is. The climate became correspondingly warm, and the icebergs melted away. Then the land rose again, and in time the inland sea was drained away into the ocean, except for the waters that remained in thousands of lakes great and small that now occupy the region covered by the ice.

Ancient sea beaches mark the level of high water at the time that the flood followed the melting ice. On the shores of Lake Champlain, but nearly five hundred feet higher than the present level of the lake, curious geologists have found many kinds of marine shells on a well-marked old sea beach. The members of one exploring party in the same region were surprised and delighted to come by digging upon the skeleton of a whale that had drifted ashore in the ancient days when the inland sea joined the Atlantic.

Lake Ontario's ancient beach is five hundred feet above the present water-level; Lake Erie's is two hundred fifty feet above it; Lake Superior's three hundred thirty feet higher than the present beach. No doubt when the water stood at the highest level, the Great Lakes formed one single sheet of water which settled to a lower level as the rivers flowing south cut their channels deep enough to draw off the water toward the Gulf. Lake Winnipeg is now the small remnant of a vast lake the shores of which have been traced. The Minnesota River finally made its way into the Mississippi and drained this great area the stranded beaches of which still remain. The name of Agassiz has been given to the ancient lake formed by the glacial flood and drained away thousands of years ago but not until it had built the terraced beach which locates it on the geological map of the region.

When the ice-sheet came down from the north it dragged along all of the soil and loose rock material that lay in its path. With the boulders frozen into its lower surface it scratched and grooved the firm bedrock over which it slid, and rounded it to a smooth and billowy surface. The progress of the ice-sheet was southward, but it spread like a fan so that its widening border turned to east and west.

When it reached its southernmost limit and began to melt, it laid down a great ridge of unsorted rock material, remnants of which remain to this day,—the terminal moraine of the ancient ice-sheet. The line of this ancient deposit starts on Long Island, crosses New Jersey and Pennsylvania, then dips southward, following the general course of the Ohio River to its mouth, forming bluffs in southern Ohio, Indiana, and Illinois. The line bends upward as it crosses central Missouri, a corner of Kansas, and eastern Nebraska, parallel with the course of the Missouri.

As the ice-sheet melted, boulders were dropped all over the Northern States and Canada. These were both angular and rounded. In some places they are scattered thickly over the surface and are so numerous as to be a great hindrance to agriculture. In many places great boulders of thousands of tons weight are perched on very slight foundations, just where they lodged when the ice went off and left them, after carrying them hundreds of miles. Around them are scattered quantities of loose rock material, not scored or ground as are those which were carried on the under-surface of the glacial ice. These unscarred fragments rode on the top of the ice. They were a part of the top moraine of the glacial sheet.

The finest material deposited is rock meal, ground by the great glacial mill, and called "boulder clay." It is a stiff, dense, stony paste in which boulders of all sizes, gravel, pebbles, and cobblestones are cemented.

The "drift" of the ice-sheet is the rubbish, coarse and fine, it left behind as it retreated. Below the Ohio River there is a deep soil produced by the decay of rocks that lie under it. North of Ohio is spread that peculiar mixture of earth and rock fragments which was transported from the north and spread over the land which the ice-sheet swept bare and ground smooth and polished.

The drift has been washed away in places by the floods that followed the ice. Granite domes are thus exposed, the grooves and scratches of which tell in what direction the ice flood was travelling. Miles away from that scored granite, but in the same direction as the scratches, scattered fragments of the same foundation rock cover fields and meadows. Thus, much of the drift material can be traced to its original home, and the course of the ice-sheet can be determined. Many immense boulders the home of which was in the northern highlands of Canada rode southward, frozen into icebergs that floated in the great inland sea. Great quantities of débris were added to the original glacial drift through the agency of these floating ice masses, which melted by slow degrees.

FOLLOWING SOME LOST RIVERS

What would you think if the boat in which you were floating down a pleasant river should suddenly grate upon sand, and you should look over the gunwale and find that here the waters sank out of sight, the river ended? I believe you would rub your eyes, and feel sure that you were dreaming. Do not all rivers flow along their beds, growing larger with every mile, and finally empty their waters into a sea, or bay, or lake, or flow into some larger stream? This is the way of most rivers, but there are exceptions. In the Far West there are some great rivers that absolutely disappear before they reach a larger body of water. They simply sink away into the sand, and sometimes reappear to finish their courses after flowing underground for miles. Do you know the name of one great western river of which I am thinking? Is there any stream in your neighbourhood which has such peculiar ways?

Down in Kentucky there is a region where, it is said, one may walk fifty miles without crossing running water. In the middle of our country, in the region of plentiful rainfall, and in a state covered with beautiful woodlands and famous for blue grass and other grain crops, it is amazing that, over a large area, brooks and larger streams are lacking. In most of the state there is plenty of water flowing in streams like those in other parts of the eastern half of the United States. In the near neighbourhood of this peculiar section of the state the streams come to an end suddenly, pouring their water into funnel-shaped depressions of the ground called sink-holes. After a heavy rain the surface water, accumulating in rivulets, may also be traced to small depressions which seem like leaks in the earth's crust, into which the water trickles and disappears.

It must have been noticed by the early settlers who came over the mountains from the eastern colonies, and settled in the new, wild, hilly country, which they called Kentucky. The first settlers built their log cabins along the streams they found, and shot deer and wild turkey and other game that was plentiful in the woods. The deer showed them where salt was to be found in earthy deposits near the streams; for salt is necessary to every creature. Deer trails led from many directions to the "salt licks" which the wild animals visited frequently.

Perhaps the same pioneers who dug the salt out of the earth found likewise deposits of *nitre*, called also *saltpetre*, a very precious mineral, for it is one of the elements necessary in the manufacture of gunpowder. With the Indians all about him, and often showing themselves unfriendly, the pioneer counted gunpowder a necessity of life. He relied on his gun to defend and to feed his family. There were men among those first settlers who knew how to make gunpowder, and saltpetre was one of the things that had to be carried across the mountains into Kentucky, until they found it in the hills. No wonder that prospectors went about looking for nitre beds in the overhanging ledges of rocks along stream-beds. In such situations the deposits of nitre were found. The earth was washed in troughs of running water to remove the clayey impurity. After a filtering through wood-ashes, the water which held the nitre in solution was boiled down, and left to evaporate, after which the crystals of saltpetre remained.

Solid masses of saltpetre weighing hundreds of pounds were sometimes found in protected corners under shelving rocks. It was no doubt in the fascinating hunt for lumps of this pure nitre that the early prospectors discovered that the streams which disappeared into the sink-holes made their way into caverns underground. Digging in the sides of ravines often made the earthy wall cave in, and the surprised prospector stood at the door of a cavern. The discoverer of a cave had hopes that by entering he might find nitre beds richer than those he could uncover on the surface, and this often turned out to be true. The hope of finding precious metals and beds of iron ore also encouraged the exploration of these caves. By the time the war of 1812 was declared, the mining of saltpetre was a good-sized industry in Kentucky. Most of the mineral was taken out of small caves, and shipped, when purified, over the mountains, on mule-back by trails, and in carts over good roads that were built on purpose to bring this mineral product to market. As long as war threatened the country, the Government was ready to buy all the saltpetre the Kentucky frontiersmen could produce. And the miners were constantly in search of richer beds that promised better returns for their labour.

It was this search that led to the exploration of the caves discovered, although the explorer took his life in his hands when he left the daylight behind him and plunged into the under-world.

Not all lost rivers tell as interesting stories or reveal as valuable secrets as did those the neighbours of Daniel Boone traced along their dark passages underground, and finally saw emerge as hillside springs, in many cases, to feed Kentucky rivers. But it is plain that no river sinks from sight unless it finds porous or honeycombed rocks that let it through. The water seeks the nearest and easiest route to the sea. Its weight presses toward the lowest level, always. The more water absorbs of acid, the more powerfully does it attack and carry away the substance of lime rocks through which it passes.

THE MAMMOTH CAVE OF KENTUCKY

There is no more fertile soil in the country than that of the famous blue grass region of Kentucky. The surface soil rests upon a deep foundation of limestone rocks, and very gradually the plant food locked up in these underlying strata is pulled up to the surface by the soil water, and greedily appropriated by the roots of the plants.

Part of the water of the abundant rainfall of this region soaks into the layers of the lime rock, carrying various acids in solution which give it power to dissolve the limestone particles, and thus to make its way easily through comparatively porous rock to the very depths of the earth. So it has come about that the surface of the earth is undermined. Vast empty chambers have been carved by the patient work of trickling water, which has carried away the lime that once formed solid and continuous layers of the earth's crust. We must believe that the work has taken thousands of years, at least, for no perceptible change has come to these wonderful caves since the discovery and exploration of them a century and more ago.

The streams that flow into the region of these caves disappear suddenly into sink-holes and flow through caverns. After wearing away their subterranean channels, leaping down from one level to another, forming waterfalls and lakes, some emerge finally through hillsides in the form of springs.

The cavern region of Kentucky covers eight thousand square miles. The underground chambers found there are in the limestone rock which varies from ten to four hundred feet in thickness, and averages a little less than two hundred feet. Over this territory the number of sink-holes average one hundred to the square mile; and the streams that have poured their water into these basins have made a network of open caverns one hundred thousand miles in length.

A great many small caverns have been thoroughly explored and are famous for their beauty. The Diamond Cave is one of the most splendid, for it is lined with walls and pillars of alabaster that sparkle in the torchlight with crystals that look like veritable diamonds. Beautiful springs and waterfalls are found in many caves, but the grandest of all is the Mammoth Cave, beside which no other is counted worthy to be compared.

Great tales the miners told of the wonder and the beauty of these caverns, the walls of which were supported by arching alabaster columns and wonderful domes, of indescribable beauty of form and colouring. In 1799, the year that Washington died, a pioneer discovered the entrance to a cave, the size and beauty of which surpassed anything he had seen before. After exploring it for a short distance he returned home and took his whole family with him to enjoy the first view of the wonderful cavern he had discovered. They carried pine knots and a lighted torch, by which they made their way for some distance, but the torch was accidentally extinguished and they groped their way in darkness and missed the entrance. Without anything to guide them, they wandered in darkness for three days, and were almost dead when at last they stumbled upon the exit. This is the doorway of the Mammoth Cave of Kentucky, one of the wonders of the world.

This was a terrible experience. The next persons who attempted to explore the new cave were better provisioned against the chance of spending some time underground. The pioneers found rich deposits of nitre in the "Great Cave," as they called it. Scientists visited it and explored many of its chambers. The reputation of this cavern has been spread by thousands of visitors who have come from all over the world to see it. The cave has not yet been completely explored. The regular tours, on which the guides conduct visitors, cover but a small part of the one hundred and fifty miles measured by the two hundred or more avenues. The passages wind in and out, crossing each other, sometimes at different levels, and forming a network of avenues in which the unaccustomed traveller would surely be lost. The old guides know every inch of their regular course, and their quaint and edifying talk adds greatly to the pleasure of the visitors.

From the hotel, parties are organized for ten o'clock in the morning and seven o'clock in the evening. Each visitor is provided with a lard-oil lamp. The guide carries a flask of oil and plenty of matches. No special garb is necessary, though people usually dress for comfort, and wear easy shoes. The temperature of the cave is uniform winter and summer, varying between fifty-three and fifty-four degrees Fahrenheit.

The cave entrance is an arch of seventy-foot span in the hillside. A winding flight of seventy stone steps leads the party around a waterfall, into a great chamber under the rocks. Then the way goes through a narrow passage, where the guide unlocks an iron gate to let them in. The visitors now leave all thoughts of daylight behind, for the breeze that put out their lights as they entered the cave is past, and they stand in the Rotunda, a vast high-ceilinged chamber, silent and impressive, with walls of creamy limestone, encrusted with gypsum, which has been stained black by manganese. From the vestibule on, each passage and each room has a name, based upon some historic event or some fancied resemblance. The Giant's Coffin is a great kite-shaped rock lying in one of the rooms of the cave. The Star Chamber has a wonderful crystal-studded dome in which the guide produces the effect of a sunrise by burning coloured lights. Bonfires built at suitable points produce wonderful shadow effects, which are like nothing else in the world. The old saltpetre vats which the visitors pass in taking the "Long Route" through the cave, point them back to the days during the War of 1812, when this valuable mineral was extracted from the earth in the floor of the cave. The industry greatly enriched the thrifty owners of the cave, but the works were abandoned after peace was declared.

It must be a wonderful experience to walk steadily for nine hours over the Long Route, for so pure is the air and so wonderful is the scenery that people rarely complain of fatigue when the experience is over. There is no dust on the floors of these subterranean chambers, and they are not damp except near places where water trickles, here and there, in rivulets and cascades. Pools of water at the bottoms of pits so deep that a lighted torch requires several seconds to reach the bottom, and rivers and lakes of considerable size, show where some of the surface water goes to. A strange underground suction creates whirlpools in some of these streams. People go in boats holding twenty passengers for a row on Echo River, and the guide dips up with a net the blind fish and crayfish and cave lizards which inhabit these subterranean waters. The echoes in various chambers of the Mammoth Cave are remarkable. In some of them a song by a single voice comes back with full chords, as if several voices carried the different parts. The single notes of flute and cornet are returned with the same beautiful harmonies. A pistol shot is given back a dozen times, the sound rebounding like a ball from rock to rock of the arching walls. The vibrations of the water made by the rower's paddles reëcho in sounds like bell notes, and they are multiplied into harmonies that suggest the chimes in the belfry of a cathedral.

The walls of various chambers differ from each other according to the minerals that compose them. Some are creamy white limestone arches, some are walled with black gypsum, some are hung with great curtains of stalagmites, solid but suggesting the lightness and grace of folds of crêpe. Under such hangings the floor is built up in stalactites. The mineral-laden water, the constant drip of which has produced a hanging, icicle-like stalagmite, has built up the stalactite to meet it.

Probably nothing is more beautiful than the flower-like crystals that bloom all over the walls of a chamber called "Mary's Bower." The floor, even, sparkles with jewels that have fallen from the wonderful and delicate flower clusters built from deposits of the lime-laden water which goes on building and replacing the bits that fall. "Martha's Vineyard" is decorated with nodules, like bunches of grapes, that glisten as if the dew were on them. The white gypsum in some caves makes the walls look as if they were carved out of snow. Still others have clear, transparent crystals that make them gleam in the torches' light as if the walls were encrusted with diamonds.

The cave region of Indiana is also famous. The great Wyandotte Cave in Crawford County is the most noted of many similar caverns. In some of the chambers, bats are found clinging to the ceiling, heads downward, like swarms of bees. The caverns of Luray, in Virginia, are complex and wonderful in their structure, and famous for the beautiful stalactites and stalagmites they contain. But there is no cave in this country so wonderful and so grand in its dimensions as the Mammoth Cave in Kentucky.

LAND-BUILDING BY RIVERS

Once a year, when the rainy season comes in the mountainous country south of Egypt, the old Nile floods its banks and spreads its slimy waters over the land, covering the low plains to the very edge of the Sahara Desert. The people know it is coming, and are prepared for this flood. We should think such an overflow of our nearest river a monstrous calamity, but the Egyptians bless the river which blesses them. They know that without the Nile's overflow their country would be added to the Desert of Sahara. In a short time after the overflow, the river reaches its highest point and begins to ebb. Canals lying parallel to its course are filled with water which is saved for use in the hot, dry summer. As the flood goes down, a deposit of slimy mud lies as a rich fertilizer on the land. It is this and the water which the earth has absorbed that make Egypt one of the most fertile agricultural countries in the world.

The region covered by the Nile's overflow is the flood plain of this river. On this plain the Pyramids, the Sphinx, and other famous monuments of Egypt stand. The statue of Rameses II. built 3,000 years ago, has its base buried nine feet deep in the rich soil made of Nile sediment. A well dug in this region goes through forty feet of this soil before striking the underlying sand. How many years ago did the first Nile overflow take place? We may begin our calculation by finding out the average yearly deposit. It is a slow process that accumulates but nine feet in 3,000 years. If you were in Egypt when the Nile went back into its banks, you would see that the scum it leaves in a single overflow adds not a great deal to the thickness of the soil. Possibly floods have varied in their deposits from year to year, so that any calculation of the time it took to build that forty feet of surface soil must be but a rough estimate. This much we know: it has been an uninterrupted process which has taken place within the present geological epoch, "the Age of Man."

Not all the rich sediment the Nile brings down is left on the level flood plain along its course. A vast quantity is dumped at the river's mouth, where the tides of the Mediterranean check the river's current. Thus the great delta is formed. The broad river splits into many mouths that spread out like a fan and build higher and broader each year the mud-banks between the streams. Upper Egypt consists of river swamps. Lower Egypt, from Cairo to the sea, is the delta built by the river itself on sea bottom. From the head of the delta, where the river commences to divide, to the sea, is an area of 10,000 square miles made out of material contributed by upper Egypt, and built by the river. Layer upon layer, it is constantly forming, but most rapidly during the season of floods.

Coming closer home, let us look at the map of the Mississippi Valley. Begin as far north as St. Louis. For the rest of its course the Mississippi River flows through a widening plain of swamp land, flooded in rainy seasons. Through this swampy flood plain the river meanders; its current, heavily loaded with sediment, swings from one side to the other of the channel, building up here, wearing away there, and straightening its course when the curves become so sharp that their sides meet. Then the current breaks through the thin wall, and a bayou of still water is left behind.

Below Baton Rouge the Mississippi breaks into many mouths, that spread and carry the water of the great river into the Gulf of Mexico. The Nile delta is triangular, like delta, [Greek: D], the fourth letter of the Greek alphabet; but the Mississippi's delta is very irregular. The main mouth of the river flows fifty miles out into the Gulf between mud-banks, narrow and low. At the tip it branches into several streams.

From the mouth of the Ohio to the Gulf, the Mississippi flood plain covers 30,000 square miles. Over this area, sediment to an average depth of fifty feet has been laid down. In earlier times the river flooded this whole area, when freshets swelled its tributaries in the spring. The flood plain then became a sea, in the middle of which the river's current flowed swiftly. The slow-flowing water on each side of the main current let go of its burden of sediment and formed a double ridge. Between these two natural walls the main river flowed. When its level fell, two side streams, running parallel with the main river drained the flood plains on each side into the main tributaries to right and left. These natural walls deposited when the river was in flood are called *levees*. Each heavy flood builds them higher, and the bed of the stream rises by deposits of sediment. So it happens that the level of the river bed is higher than the level of its flood plain.

This is an interesting fact in geology. But the people who have taken possession of the rich flood plain of the Mississippi River, who have built their homes there, drained and cultivated the land, and built cities and towns on the areas reclaimed from swamps, recognize the elevation of the river bed as the greatest danger that threatens them. Suppose a flood should come. Even if it does not overflow the levees, it may break through the natural banks and thus overflow the cities and the farm lands to left and right.

Instead of living in constant fear of such a calamity, the people of the Mississippi flood plain have sought safety by making artificial levees, to make floods impossible. These are built upon the natural levees. As the river bed rises by the deposit of mud, the levees are built higher to contain the rising waters. No longer does the rich soil of the Mississippi flood plain receive layers of sediment from the river's overflow. The river very rarely breaks through a levee. The United States Government has spent great sums in walling in the river, and each state along its banks does its share toward paying for this self-protection.

By means of *jetties* the river's current is directed into a straightened course, and its power is expended upon the work of deepening its own channel and carrying its sediment to the Gulf. Much as the river has been forced to do in cleaning its own main channel, dredging is needed at various harbours to keep the river deep enough for navigation. The forests of the mountain slopes in Colorado are being slaughtered, and the headwaters of the Missouri are carrying more and more rocky débris to choke the current of the Mississippi. Colorado soil is stolen to build land in the vast delta, which is pushing out into the Gulf at the rate of six miles in a century—a mile in every sixteen years. The Mississippi delta measures 14,000 square miles. With the continued denuding of mountain slopes, we shall expect the rate of delta growth to be greatly increased, until reforesting checks the destructive work of wind and water.

THE MAKING OF MOUNTAINS

The gradual thickening and shrinking of the earth's crust as it cools have made the wrinkles we call mountain systems. Through millions of years the globe has been giving off heat to the cold sky spaces through which it swings in its orbit around the sun. The cooling caused the contraction of the outer layer to fit the shrinking of the mass. When a plump peach dries on its pit, the skin wrinkles down to fit the dried flesh. The fruit shrinks by loss of water, just as the face of an old person shrinks by loss of fat. The skin becomes wrinkled in both cases.

The weakest places in the earth's crust were the places to crumple, because they could not resist the lateral pressure that was exerted by the shrinking process. Along the shores of the ancient seas the rivers piled great burdens of sediment. This caused the thin crust to sink and to become a basin alongside of a ridge. The wearing away of the land in certain places lightened and weakened the crust at these places, so that it bent upward in a ridge.

Perhaps the first wrinkles were not very high and deep. The gradual cooling must have exerted continued pressure, and the wrinkles have become larger. It is not likely that new wrinkles would be formed as long as the old ones would crumple and draw up into narrower, steeper slopes, in response to the lateral crushing.

We can imagine those first mountains rising as folds under the sea. Gradually their bases were narrowed, and their crests lifted out of the water. They rose as long, narrow islands, and grew in size as time went on.

Why is the trend of the great mountain systems almost always north and south? Study the map of the continents and see how few cross ranges are shown, and how short they are, compared with the others. The molten globe bulged at its equator, as it rotated on its axis. The moon added its strong pulling force to make it bulge still more. As the crust thickened, it became less responsive to the two forces that caused it to bulge. The shrinkage was greatest where the globe had been most pulled out of shape. The rate of the earth's rotation is believed to have diminished. Every change tended to let the earth draw in its (imaginary) belt, a notch at a time. The forces of contraction acted along the line of the equator, and formed folds running toward the poles. In this early time the great mountain systems were born, and they grew in size gradually, from small beginnings.

These mountains of upheaval, made by the bending of the earth's crust, and the formation of alternating ridges and depressed valleys, are many. The earth is old and much wrinkled. Other mountains have been formed by forces quite different. Volcanic mountains have been far more numerous in ages gone than they are now.

Mt. Hood and Mt. Rainier are peaks built up by the materials thrown out of the craters of volcanoes dead these thousands of years. Vesuvius is at present showing us how volcanic mountains are made. Each eruption builds larger the cone—that is, the chimney through which the molten rocks, the ashes, and the steam are ejected. Side craters may open, the main cone be broken and its form changed, but the mass of lava and stones and ashes grows with each eruption. The mountain grows by the additions it receives. Ætna is a mountain built of lava.

A third mountain system grew, not by addition, but by subtraction. The Catskills illustrate this type. This group of mountains is the remnant of a table-land made of level layers of red sandstone. The rest of the high plain has been cut down and carried away, leaving these picturesque hills, the survival of which is as much a mystery as the disappearance of the balance of the plateau of which they were once a part.

The fold that forms a typical mountain ridge has a cone of granite, the original rock foundation of the earth, and on this are layers of stratified rock, ancient deposits of sediment carried to the sea by streams. When exposed to wind and rain, the ridge is gradually worn down. In some places the water cuts away the soft rock and forms a stream-bed, that cuts deeper and deeper, using the rock fragments as its tools. Often the layers of aqueous rocks are cut through, and the granite exposed.

Sometimes the hardest stratified rock-beds resist the water and the wind and are left as a series of ridges along the sides of the main range. The crumpling forces may crack the ridge open for its whole length, and one side of the chasm may slip down and the other go up. The result is a sheer wall of exposed rock strata, layers of which correspond with those that lie far below the top of the portion that slid down in the great upheaval and subsidence that parted them. These slips are known as *faults*.

THE LAVA FLOOD OF THE NORTHWEST

We know little about the substance that occupies the four thousand miles of distance between the surface and the centre of our earth. We know that the terrible weight borne by the central mass compresses it, so that the interior must grow denser as the core is approached. Scientists have weighed the earth, and tell us that the crust is lighter than the rest. The supposition is that there is a great deal of iron in the interior, and possibly precious metals, too.

Our deepest wells and mines go down about a mile, then digging stops, on account of the excessive heat. But the crumpling of the crust, and the wearing away of the folded strata by wind and running water, have laid bare rocks several miles in thickness on the slopes of mountains, and exposed the underlying granite, on which the first sedimentary rocks were deposited. On this granite lie stratified rocks, which are crystalline in texture. These are the beds, sometimes miles in depth, called *metamorphic* rocks, formed by water, then transformed by heat.

The wearing away of rocks by wind and water has furnished the materials out of which the aqueous rocks have been made. Layers upon layers of sandstone, shales, limestone, and the like, are exposed when a river cuts a canyon through a plateau. The layered deposits of débris at the mouth of the river make new aqueous rocks out of old. Every sandy beach is sandstone in the making. This work is never ended.

In the early days the earth's crust often gaped open in a mighty crater and let a flood of lava overspread the surface. The ocean floor often received this flood of melted rock. In many places the same chimney opened again and again, each time spreading a new layer of lava on top of the old, so that the surface has several lava sheets overlying the aqueous strata.

If the hardened lava sheet proved a barrier to the rising tide of molten lava in the chimney it was often forced out in sheets between the layers of aqueous rocks. Wherever the heated material came into contact with aqueous rocks it transformed them, for a foot or more, into crystalline, metamorphic rocks.

A chimney of lava is called a *dike*. In mountainous countries dikes are common. Sometimes small, they may also be hundreds of feet across, often standing high above the softer strata, which rains have worn away. Dikes often look like ruined walls, and may be traced for miles where they have been overturned in the mountain-making process.

The great lava flood of the Northwest happened when the Coast Range was born. Along the border of the Pacific Ocean vast sedimentary deposits had accumulated during the Cretaceous and Tertiary Periods. Then the mighty upheaval came, the mountain ridge rose at the end of the Miocene epoch and stretched itself for hundreds of miles through the region which is now the coast of California and Oregon. Great fissures opened in the folded crust, and floods of lava overspread an area of 150,000 square miles. A dozen dead craters show to-day where those immense volcanic chimneys were. The depth of the lava-beds is well shown where the Columbia River has worn its channel through. Walls of lava three thousand feet in thickness rise on each side of the river. They are made of columns of basalt, fitted together, like cells of a honeycomb, and jointed, forming stone blocks laid one upon another. The lava shrinks on cooling and forms prisms. In Ireland, the Giants' Causeway is a famous example of basaltic formation. In Oregon, the walls of the Des Chutes River show thirty lava layers, each made of vertical basalt columns. The palisades of the Hudson, Mt. Tom, and Mt. Holyoke are examples on the eastern side of the continent of basaltic rocks made by lava floods.

Northern California, northwestern Nevada, and large part of Idaho, Montana, Oregon, and Washington are included in the basin filled with lava at the time of the great overflow, which extended far into British Columbia. It is probable that certain chimneys continued to discharge until comparatively recent times. Mt. Rainier, Mt. Shasta, and Mt. Hood are among dead volcanoes.

Quite a different history has the great Deccan lava-field of India, which covers a larger area than the basin of our Northwest, and is in places more than a mile in depth. It has no volcanoes, nor signs of any ever having existed. The floods alone overspread the region, which shows no puny "follow-up system" of scattered craters, intermittently in eruption.

THE FIRST LIVING THINGS

Strange days and nights those must have been on the earth when the great sea was still too hot for living things to exist in it. The land above the water-line was bare rocks. These were rapidly being crumbled by the action of the air, which was not the mild, pleasant air we know, but was full of destructive gases, breathed out through cracks in the thin crust of the earth from the heated mass below. If you stand on the edge of a lava lake, like one of those on the islands of the Hawaiian group, the stifling fumes that rise might make you feel as if you were back at the beginning of the earth's history, when the solid crust was just a thin film on an unstable sea of molten rock, and this volcano but one of the vast number of openings by which the boiling lava and the condensed gases found their way to the surface. Then the rivers ran black with the waste of the rocky earth they furrowed, and there was no vegetation to soften the bleakness of the landscape.

The beginnings of life on the earth are a mystery. Nobody can guess the riddle. The earliest rocks were subjected to great heat. It is not possible that life could have existed in the heated ocean or on the land. Gradually the shores of the seas became filled up with sediment washed down by the rivers. Layer on layer of this sediment accumulated, and it was crumpled by pressure, and changed by heat, so that if any plants or animals had lived along those old shores their remains would have been utterly destroyed.

Rocks that lie in layers on top of these oldest, fire-scarred foundations of the earth show the first faint traces of living things. Limestone and beds of iron ore are signs of the presence of life. The first animals and plants lived in the ancient seas.

From the traces that are left, we judge that the earliest life forms were of the simplest kind, like some plants and animals that swim in a drop of water. Have you ever seen a drop of pond water under a compound microscope? It is a wonder world you look into, and you forget all the world besides. You are one of the wonderful higher animals, the highest on the earth. You focus on a shapeless creature that moves about and feels and breathes, but hasn't any eyes or mouth or stomach—in fact, it is the lowest form of animal life, and one of the smallest. It is but one of many animal forms, all simple in structure, but able to feed and grow and reproduce their kind.

Gaze out of the window on the garden, now. The flowering plants, the green grass, and the trees are among the highest forms of plants. In the drop of water under the microscope tiny specks of green are floating. They belong to the lowest order of plants. Among the plant and the animal forms that have been studied and named, are a few living things the places of which in the scale are not agreed upon. Some say they are animals; some believe they are plants. They are like both in some respects. It is probable that the first living things were like these confusing, minute things—not distinctly plants or animals, but the parent forms from which, later on, both plants and animals sprang.

The lowest forms of life, plant and animal, live in water to-day. They are tiny and their bodies are made of a soft substance like the white of an egg. If these are at all like the living creatures that swarmed in the early seas, no wonder they left no traces in the rocks of the early part of the age when life is first recorded by fossils. Soft-bodied creatures never do.

Some of the animals and the plants in the drop of water under the microscope have body walls of definite shapes, made of lime, or of a glassy substance called silica. When they die, these "skeletons" lie at the bottom of the water, and do not decay, as the living part of the body does, because they are mineral. Gradually a number of these shells, or hard skeletons, accumulate. In a glass of pond water they are found at the bottom, amongst the sediment. In a pond how many thousands of these creatures must live and their shells fall to the bottom at last, buried in the mud!

So it is easy to understand why the first creatures on earth left no trace. The first real fossils found in the rocks are the hard shells or skeletons of the first plants and animals that had hard parts.

AN ANCIENT BEACH AT EBB TIDE

When the tide is out, the rocks on the Maine coast have plenty of living creatures to prove this northern shore inhabited. Starfishes lurk in the hollows, and the tent-shaped shells of the little periwinkle encrust the wet rocks. Mussels cling to the rocks in clumps, fastened to each other by their ropes of coarse black hair. The furry coating of sea mosses that encrust the rocks is a hiding-place for many kinds of living things, some soft-bodied, some protected by shells. The shallow water is the home of plants and animals of many different kinds. As proof of this one finds dead shells and fragments of seaweeds strewn on the shore after a storm.

Along the outer shores of the Cape Cod peninsula and down the Jersey coast, the sober colouring of the shells of the north gives way to a brighter colour scheme. In the warmer waters, life becomes gayer, if we may judge by the rich tints that ornament the shells. The kinds of living creatures change. They are larger and more abundant. The seaweeds are more varied and more luxuriant in growth.

When we reach the shores of the West Indian islands and the keys of Florida the greatest abundance and variety of living forms are found. The submerged rocks blossom with flower-like sea anemones of every colour. Corals, branching like trees and bushes on the sea floor, form groves under water. Among them brilliant-hued fishes swim, and highly ornamented shells glide, as people know who have gazed through the glass bottoms of the boats built especially to show visitors the wonderful sea gardens at Nassau, Bahama Islands, and at Santa Catalina Island, southern California.

On every beach the skeletons of animals which die help to build up the land; though the process is not so rapid in the north as on the shores that approach the tropics. The coast of Florida has a rim of island reefs around it built out of coral limestone. Indeed, the peninsula was built by coral polyps. Houses in St. Augustine are built of coquina rock, which is simply a mass of broken shells held together by a lime cement. Every sea beach is packed with shells and other remnants of animals and plants that live in the shallow waters. Deeper and deeper year by year the sand buries these skeletons, and many of them are preserved for all time.

Thus what is sandy beach to-day may, a few million years from now, be uncovered as a ledge of sandstone with the prints of waves distinctly shown, and fossil shells of molluscs, skeletons of fishes, and branches of seaweed—all of them different from those then growing upon the earth.

In the neighbourhood of Cincinnati there have been uncovered banks of stone accumulated along the border of an ancient sea. From the sides of granite highlands streams brought down the sand built into these oldest sandstone rocks. The fine mud which now appears as beds of slate was the decay of feldspar and hornblende in the same granite. Limestone beds are full of the fossil shells of creatures that lived in the shallow seas. Their skeletons, accumulating on the bottom, formed deep layers of limestone mud. These rocks preserve, by the fossils they contain, a great variety of shellfish which had limy skeletons. The sea fairly swarmed along its shallow margin with these creatures. We might not recognize many of the shells and other curious fossils we find in the rock uncovered by the workmen who are cutting a railroad embankment. They are not exactly like the living forms that grow along our beaches to-day, but they are enough like them for us to know that they lived along the seashore, and if we had time to examine all the rocks of this kind preserved in a museum we should decide that seashore life was quite as abundant then as it is now. The pressed specimens of plants of those earliest seashores are mere imprints showing that they were pulpy and therefore gradually decayed. Only their shape is recorded by dark stains made by each branching part. The decay of the vegetable tissue painted the outline on the rock which when split apart shows us what those ancient seaweeds looked like. They belonged to the group of plants we call kelp, or tangle, which are still common enough in the sea, though the forms we now have are not exactly like the old ones. Seaweeds belong to the very lowest forms of plants.

Crinoid from Indiana

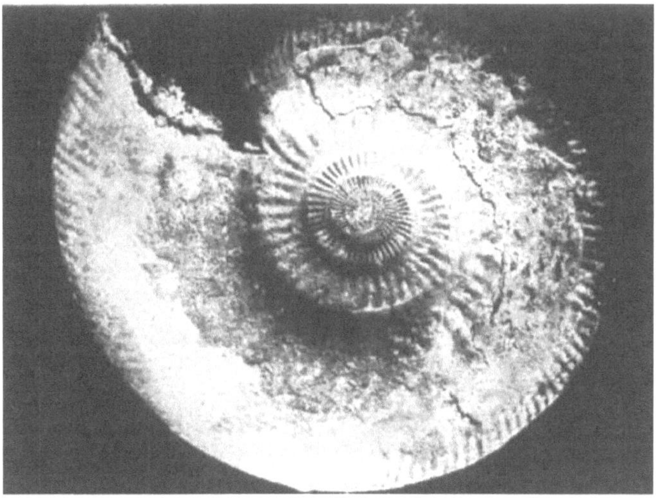

By permission of the American Museum of Natural History
Ammonite from Jurassic of England

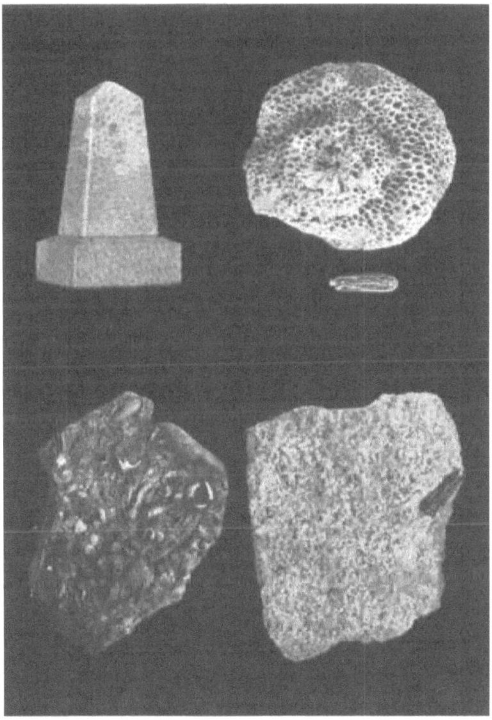

By permission of the American Museum of Natural History
Fossil corals Coquina, Hippurite limestone

The limestones are full of fossils of corals. Indeed, there must have been reefs like those that skirt Florida to-day built by these lime-building polyps. Their forms are so well preserved in the rocks that it is possible to know just how they looked when they grew in the shallows.

One very common kind is called a cup coral, because the polyp formed a skeleton shaped like a cup. The body wall surrounded the skeleton, and the arms or tentacles rose from the centre of the funnel-like depression in the top. Little cups budded off from their parents, but remained attached, and at length the skeletons of all formed great masses of limy rock. Some cup corals grew in a solid mass, the new generation forming an outer layer, thus burying the parent cups.

A second type of corals in these oldest limestones is the honeycomb group. The colonies of polyps lived in tubes which lengthened gradually, forming compact, limy cylinders like organ pipes, fitted close together. The living generation always inhabited the upper chambers of the tubes. A third type is the chain coral, made of tubes joined in rows, single file like pickets of a fence. But these walls bend into curious patterns, so that the cross-section of a mass of them looks like a complex pattern of crochet-work, the irregular spaces fenced with chain stitches. Each open link is a pit in which a polyp lived.

Among the corals are sprays of a fine feathery growth embedded in the limestone. Fine, straight, splinter-like branches are saw-toothed on one or both edges. These limy fossils might not be seen at all, were they not bedded in shales, which are very fine-grained. Here again are the skeletons of animals. Each notch on each thread-like branch was the home of a tiny animal, not unlike a sea anemone and a coral polyp.

To believe this story it is necessary only to pick up a bit of dead shell or floating driftwood on which a feathery growth is found. These plumes, like "sea mosses," as they are called, are not plants at all, but colonies of polyps. Each one lived in a tiny pit, and these pits range one above the other, so as to look like notches on the thread-like divisions of the stem. Put a piece of this so-called "sea moss" in a glass of sea water, and in a few moments of quiet you will see, by the use of a magnifying glass, the spreading arms of the polyp thrust out of each pit.

The ancient seas swarmed with these living hydrozoans, and their remains are found preserved as fossils in the shales which once were beds of soft mud.

The hard shells of sea urchins and starfishes are made of lime. In the ancient seas, starfishes were rare and sea urchins did not exist, but all over the sea bottom grew creatures called crinoids, the soft parts of which were enclosed in limy protective cases and attached to rocks on the sea bottom by means of jointed stems. No fossils are more plentiful in the early limestones than these wonderful "stone lilies." Indeed, the crinoidal limestone seemed to be built of the skeletons of these animals. The lily-like body was flung open, as a lily opens its calyx, when the creature was feeding. But any alarm caused the tentacles to be drawn in, and the petal-like divisions of the body wall to close tightly together, till that wall looked like an unopened bud.

On the bottom of the Atlantic, near the Bahama Islands, these stone lilies are still found growing. Their jointed stems and body parts are as graceful in form and motion as any lily. The creature's mouth is in the centre of the flower-like top, and it feeds like the sea urchin, on particles obtained in the sea water.

The old limestones contain great quantities of "lamp shells," which are old-fashioned bivalves. Their shells remind us of our bivalve clams and scallops, but the internal parts were very different. The gills of clams and oysters are soft parts. Inside of the lamp shells are coiled, bony arms, supporting the fringed gills.

It is fortunate for us that a few lamp shells still live in the seas. By studying the soft parts of these living remnants of a very old race we can know the secrets of the lives of those ancient lamp shells, the soft parts of which were all washed away, and the fossil shells of which are preserved. Gradually the lamp shells died out, and the modern bivalves have come to take their places. Just so, the ancient crinoids are now almost extinct; the sea urchins and the starfishes have succeeded them.

The chambered nautilus has its shell divided by partitions and it lives in the outer chamber, a many-tentacled creature, that is a close relative of the soft-bodied squid. In the ancient seas the same family was represented by huge creatures the shells of which were chambered, but not coiled. Their abundance and great size are proved by the rocks in which their fossils are preserved. Some of them must have been the rulers of the sea, as sharks and whales are to-day. Fossil specimens have been found more than fifteen feet long and ten inches in diameter in the ancient rocks of some of the Western States. It is possible to read from the lowest rock formations upward, the rise of these sea giants and their gradual decline. Certain strata of limestone contain the last relics of this race, after which they became extinct. As the straight-chambered forms diminished, great coiled forms became more abundant, but all died out.

One of the most abundant fossil animals in ancient rocks is called a trilobite. Its body is divided by two grooves into three parts, a central ridge extending the whole length of the body and two side ridges. The front portion of the shell formed the head shield, and behind the main body part was a little tail shield. The skeleton was formed of many movable jointed plates, and the creature had eyes set in the head shield just as the king crab's are set. Jointed legs in pairs fringed each side of the body. Each leg had two branches, one for walking, the other for swimming. A pair of feelers rose from the head. The body could be rolled into a ball when danger threatened, by bringing head and tail together.

These remarkable, extinct trilobites were the first crustaceans. Their nearest living relative to-day is the horseshoe crab. The fresh-water crayfish and the lobster are more distant relatives: so are the shrimps and the prawns. No such abundance of these creatures exists to-day as existed when the trilobites thronged the shallows. So well preserved are these skeletons that, although there are no living trilobites for comparison, it is possible to find out from the fossil enough about their structure to know how they fed and lived their lives along with the straight-horns which were the scavengers of those early seas and the terror of smaller creatures. The trilobites throve, and, dying, left their record in the rocks; then disappeared entirely. We find their fossils in a great variety of forms, shapes, and sizes. The smallest is but a fraction of an inch long, the largest twenty inches long.

The ancient rocks, in which these lower forms of life have left their fossils, are known as the Silurian system. The time in which these rocks were accumulating under the seas covers a vast period. We call it the Age of Invertebrates, because these soft-bodied, hard-shelled animals, the crinoids, the molluscs, and the trilobites, with bony external skeletons and no backbones, were the most abundant. They overshadowed all other forms of life. The rocks of this wonderful series were formed on the shores of a great inland sea that covered the central portion of North America. In the ages that followed, these rocks were covered deeply with later sediments. But the upheavals of the crust have broken open and erosion has uncovered these strata in different regions. Geologists have found written there, page upon page, the record of life as it existed in the early seas.

THE LIME ROCKS

"Hard" water and "soft" water are very different. The rain that falls and fills our cisterns is not softer or more delightful to use than the well water in some favoured regions. In it, soap makes beautiful, creamy suds, and it is a real pleasure to put one's hands into it. But in hard water soap seems to curdle, and some softening agent like borax has to be added or the water will chap the hands. There is little satisfaction in using water of this kind for any purpose.

Hard water was as soft as any when it fell from the sky; but the rain water trickled into the ground and passed through rocks containing lime. Some of this mineral was absorbed, for lime is readily soluble in water. Clear though it may be, water that has lime in it has quite a different feeling from rain water. Blow the breath into a basin of hard water, and a milky appearance will be noted. The carbonic acid gas exhaled from the lungs unites with the invisible lime, causing it to become visible particles of carbonate of lime, which fall to the bottom of the basin.

Nearly all well water is hard. So is the water of lakes and rivers and the ocean, for limestone is one of the most widely distributed rocks in the surface of the earth. Rain water makes its way into the earth's crust, absorbs mineral substances, and collects in springs which feed brooks and rivers and lakes. Wells are holes in the ground which bore into water-soaked strata of sand.

We gain something from the lime dissolved in hard water, for it is an essential part of our food. We must drink a certain amount of water each day to keep the body in perfect health. The lime in this water goes chiefly to the building of our bones. Plant roots take up lime in the water that mounts as sap through the plant bodies. We get some of the lime we need in vegetable foods we eat.

All of the kingdom of vertebrate animals, from the lowest forms to the highest, all of the shell-bearing animals of sea and land, require lime. Many of the lower creatures especially these in the sea, such as corals and their near relatives, encase themselves in body walls of lime. They absorb the lime from the sea water, and deposit it as unconsciously as we build the bony framework of our bodies.

All the bone and shell-bearing creatures that die on the earth and in the sea restore to the land and to the water the lime taken by the creatures while they lived. Carbonic acid gas in the water greatly hastens the dissolving of dead shells. Carbonic acid gas, whether free in the air, or absorbed by percolating water, hastens the dissolving of skeletons of creatures that die upon land. Then the raw materials are built again into lime rocks underground.

The lime rocks are the most important group in the list of rocks that form the crust of the earth. They are made of the elements calcium, carbon, and oxygen, yet the different members of this calcite group differ widely in composition and appearance. So do oyster shells and beef bones, though both contain quantities of carbonate of lime.

Calcite is a soft mineral, light in weight, sometimes white, but oftener of some other colour. It may be found crystallized or not. Whenever a drop of acid touches it, a frothy effervescence occurs. The drop of acid boils up and gives off the pungent odour of carbonic acid gas.

The reason that calcite is hard to find in rocks is that percolating water, charged with acids, is constantly stealing it, and carrying it away into the ocean. The rocks that contain it crumble because the limy portions have been dissolved out.

Some limestones resist the destructive action of water. When they are impregnated with silica they become transformed into marble, which takes a high polish like granite. Acids must be strong to make any impression on marble.

The thick beds of pure limestone that underlie the surface soil in Kentucky and in parts of Virginia sometimes measure several hundred feet in thickness, a single stratum often being twenty feet thick. They are all horizontal, for they were formed on sea bottom, and have not been crumpled in later time. The dead bodies of sea creatures contributed their shells and skeletons to the lime deposit on the sea bottom. Who can estimate the time it took to form those thick, solid layers of lime rock? The animals were corals, crinoids, and molluscs. Little sand and clay show in the lime rock of this period, before the marshes of the Carboniferous Age took the place of the ancient inland sea of the Subcarboniferous Period, the sedimentary accumulations of which we are now talking about.

The living corals one sees in the shallow water of the Florida coast to-day are building land by building up their limy skeletons. The reefs are the dead skeletons of past generations of these tiny living things. They take in lime from the water, and use it as we use lime in building our bones. In each case it is an unconscious process of animal growth—not a "building process" like a mason's building of a wall. Many people think that the coral polyp builds in this way. They give it credit for patience in a great undertaking. All the polyp does is to feed on whatever the water supplies that its digestive organs can use. It is like a sea anemone in appearance and in habits of life. It is not at all like an insect. Yet it is common to hear people speak of the "coral insect"! Do not let any one ever hear you repeat such a mistake.

Southern Florida is made out of coral rock, but thinly covered with soil. It was made by the growth of reef after reef, and it is still growing.

The Cretaceous Period of the earth's eventful history is named for the lime rock which we know as chalk. Beds of this recent kind of limestone are found in England and in France, pure white, made of the skeletons of the smallest of lime-consuming creatures, Foraminifera. They swarmed in deep water, and so did minute sponge animalcules and plant forms called Diatoms that took silica from the water, and formed their hard parts of this glassy substance. The result is seen in the nodules of flint found in the soft, snow-white chalk. Did you ever use a piece of chalk that scratched the black-board? The flint did it. Have you ever seen the chalk cliffs of Dover? When you do see them, notice how they gleam white in the sun. See how the rains have sculptured those cliffs. The prominences left standing out are strengthened by the flint they contain. Chalk beds occur in Texas and under our great plains; but the principal rocks of the age in America were sandstones and clays.

THE AGE OF FISHES

The first animal with a backbone recorded its existence among the fossils found in rocks of the upper Silurian strata. It is a fish; but the earliest fossils are very incomplete specimens. We know that these old-fashioned fishes were somewhat like the sturgeons of our rivers. Their bodies were encased in bony armour of hard scales, coated with enamel. The bones of the spine were connected by ball and socket joints, and the heads were movable. In these two particulars the fishes resembled reptiles. The modern gar-pike has a number of the same characteristics.

Another backboned creature of the ancient seas was the ancestral type of the shark family. In some points this old-fashioned shark reminds us of birds and turtles. These early fishes foreshadowed all later vertebrates, not yet on the earth. After them came the amphibians, then the reptiles, then the birds, and latest the mammals.

The race of fishes began, no doubt, with forms so soft-boned that no fossil traces are preserved in the rocks. When those with harder bones appeared, the fossil record began, and it tells the story of the passing of the early, unfish-like forms, and the coming of new kinds, great in size and in numbers, that swarmed in the seas, and were tyrants over all other living things. They conquered the giant straight-horns and trilobites, former rulers of the seas.

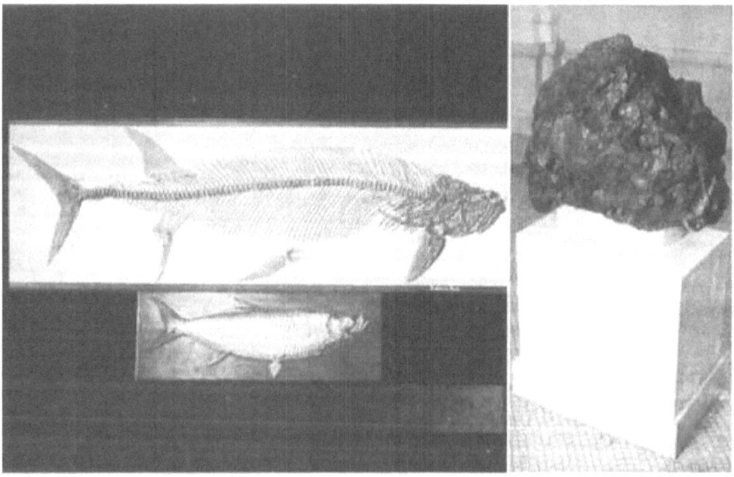

By permission of the American Museum of Natural History
A sixteen-foot fossil fish from Cretaceous of Kansas, with a modern tarpon

By permission of the American Museum of Natural History
Cañon Diablo meteorite from Arizona

One of these giant fishes fifteen to twenty feet long, three feet wide, had jaws two feet long, set with blade-like teeth. Devonian rocks in Ohio have yielded fine fossils of gigantic fishes and sharks.

Devonian fishes were unlike modern kinds in these particulars, the spinal column extended to the end of the tail, whether the fins were arranged equally or unequally on the sides; the paired side fins look like limbs fringed with fins. Every Devonian fish of the gar type seems to have had a lung to help out its gill-breathing.

In these traits the first fishes were much like the amphibians. They were the parent stock from which branched later the true fishes and the amphibians, as a single trunk parts into two main boughs. The trunk is the connecting link.

The sea bottom was still thronged with crinoids, and lamp shells, and cup corals. Shells of both clam and snail shapes are plentiful. The chambered straight-horns are fewer and smaller, and coiled forms of this type of shell are found. Trilobite forms are smaller, and their numbers decrease.

The first land plants appeared during this age. Ferns and giant club mosses and cycads grew in swampy ground. This was the beginning of the wonderful fern forests that marked the next age, when coal was formed.

The rocks that bear the record of these living things in their fossils, form strata of great thickness that overlie the Silurian deposits. There is no break between them. So we understand that the sea changed its shore-line only when the Silurian deposits rose to the water-level.

The Devonian sea was smaller than the Silurian. A great tract of Devonian deposits occupies the lower half of the state of New York, Canada between Lakes Erie and Huron, and the northern portions of Indiana and Illinois. These older layers of the stratified rock are covered with the deposits of later periods. Rivers that cut deep channels reveal the earlier rocks as outcrops along their canyon walk. The record of the age of fishes is, for the most part, still an unopened book. The pages are sealed, waiting for the geologist with his hammer to disclose the mysteries.

KING COAL

In this country, and in this age, who can doubt that coal is king? It is one of the few necessities of life. In various sections of the country, layers of coal have been discovered—some near the surface, others deep underground. These are the storehouses of fuel which the coal miners dig out and bring to the surface, and the railroads distribute. From Pennsylvania and Ohio to Alabama stretches the richest coal-basin. Illinois and Indiana contain another. From Iowa southward to Texas another broad basin lies. Central Michigan and Nova Scotia each has isolated coal-basins. All these have been discovered and mined, for they lie in the oldest part of the country.

In the West, coal-beds have been discovered in several states, but many regions have not yet been explored. Vast coal-fields, still untouched, have been located in Alaska. The Government is trying to save this fuel supply for coming generations. Many of the richest coal-beds from Nova Scotia southward dip under the ocean. They have been robbed by the erosive action of waves and running water. Glaciers have ground away their substance, and given it to the sea. Much that remains intact must be left by miners on account of the difficulties of getting out coal from tilted and contorted strata.

As a rule, the first-formed coal is the best. The Western coal-fields belong to the period following the Carboniferous Age. Although conditions were favourable to abundant coal formation, Western coal is not equal to the older, Eastern coal. It is often called *lignite*, a word that designates its immaturity compared with anthracite.

Coal formed in the Triassic Period is found in a basin near Richmond, Virginia. There is an abundance of this coal, but it has been subjected to mountain-making pressure and heat, and is extremely inflammable. The miners are in constant danger on account of coal gas, which becomes explosive when the air of the shaft reaches and mingles with it. This the miner calls "fire damp." North Carolina has coal of the same formation, that is also dangerous to mine, and very awkward to reach, on account of the crumpling of the strata.

There are beds of coal so pure that very little ash remains after the burning. Five per cent, of ash may be reasonably expected in pure coal, unmixed with sedimentary deposits. Such coal was formed in that part of the swamp which was not stirred by the inflow of a river. Wherever muddy water flowed in among the fallen stems of plants, or sand drifted over the accumulated peat, these deposits remained, to appear later and bother those who attempt to burn the coal.

Eocene fish

By permission of the American Museum of Natural History
Trilobite from the Niagara limestone, Upper Silurian, of Western New York

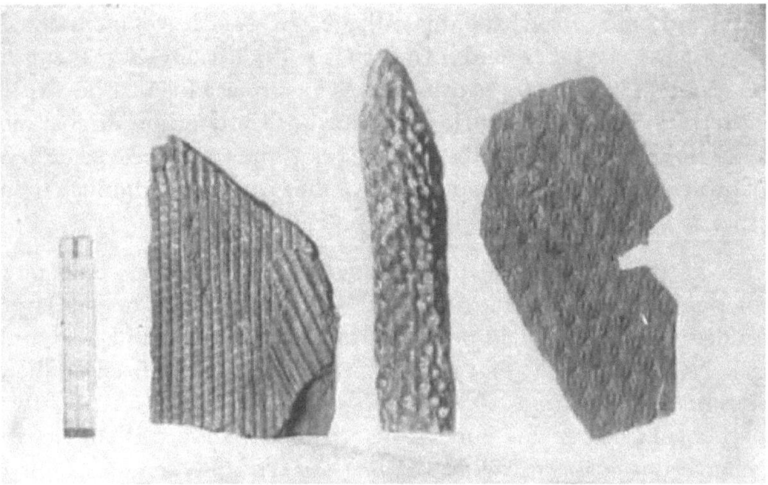

Sigillaria, Stigmaria and Lepidodendron

By permission of the American Museum of Natural History
Coal fern

You know pure coal, that burns with great heat and leaves but little ashes. You know also the other kind, that ignites with difficulty, burns with little flame, gives out little heat, and dying leaves the furnace full of ashes. You are trying to burn ancient mud that has but a small proportion of coal mixed with it. The miners know good coal from poor, and so do the coal dealers. It is not profitable to mine the impure part of the vein. It costs as much to mine and ship as the best quality, and it brings a much lower price.

The deeper beds of coal are better than those formed in comparatively recent time and found lying nearer the surface. In many bogs a layer of embedded root fibres, called peat, is cut into bricks and dried for burning. Deeper than peat-beds lie the *lignites*, which are old beds of peat, on the way to become coal. *Soft coal* is older than lignite. It contains thirty to fifty per cent. of volatile matter, and burns readily, with a bright blaze. The richest of this bituminous coal is called *fat*, or *fusing coal*. The bitumen oozes out, and the coal cakes in burning. Ordinary soft coal contains less, but still we can see the resinous bitumen frying out of it as it burns. There is more heat and less volatile matter in *steam coal*, so-called because it is the fuel that most quickly forms steam in an engine. *Hard coal* contains but five to ten per cent. of volatile matter. It is slow to ignite and burns with a small blue blaze.

From peat to anthracite coal I have named the series which increases gradually in the amount of heat it gives out, and increases and then decreases in its readiness to burn and in the brightness of its flame. Anthracite coal has the highest amount of fixed carbon. This is the reason why it makes the best fuel, for fixed carbon is the substance which holds the store of imprisoned sunlight, liberated as heat when the coal burns. Tremendous pressure and heat due to shrinking of the earth's crust have crumpled and twisted the strata containing coal in eastern Pennsylvania, and thus changed bituminous coal into anthracite. Ohio beds, formed at the same time, but undisturbed by heat and pressure, are bituminous yet.

The coal-beds of Rhode Island are anthracite, but the coal is so hard that it will not burn in an open fire. The terrible, mountain-making forces that crumpled these strata and robbed the coal of its volatile matter, left so little of the gas-forming element, that a very special treatment is necessary to make the carbon burn. It is used successfully in furnaces built for the smelting of ores.

The last stage in the coal series is a black substance which we know as black lead, or graphite. We write with it when we use a "lead" pencil. This is anthracite coal after all of the volatile matter has been driven out of it. It cannot burn, although it is solid carbon. The beds of graphite have been formed out of coal by the same changes in the earth's crust which have converted soft coal into anthracite.

The tremendous pressure that bears on the coal measures has changed a part of the carbon into liquid and gaseous form. Lakes of oil have been tapped from which jets of great force have spouted out. Such accumulations of oil usually fill porous layers of sandstone and are confined by overlying and underlying beds of impervious clay. Gas may be similarly confined until a well is drilled, relieving the pressure, and furnishing abundant or scanty supply of this valuable fuel. Western Pennsylvania coal-fields have beds of gas and oil. If mountain-making forces had broken the strata, as in eastern Pennsylvania, the gas and the oil would have been lost by evaporation.

This is what happened in the anthracite coal-belt.

HOW COAL WAS MADE

The broad, rounded dome of a maple tree shades my windows from the intense heat of this August day. The air is hot, and every leaf of the tree's thatched roof is spread to catch the sunlight. The carbon in the air is breathed in through openings on the under side of each leaf. The sap in the leaf pulp uses the carbon in making starch. The sun's heat is absorbed. It is the energy that enables the leaf-green to produce a wonderful chemical change. Out of soil water, brought up from the roots, and the carbonic acid gas, taken in from the air, rich, sugary starch is manufactured in the leaf laboratory.

This plant food returns in a slow current, feeding the growing cells under the bark, from leaf tip to root tip, throughout the growing tree. The sap builds solid wood. The maple tree has been built out of muddy water and carbon gas. It stands a miracle before our eyes. In its tough wood fibres is locked up all the heat its leaves absorbed from the sun, since the day the maple seed sprouted and the first pair of leaves lifted their palms above the ground.

If my maple tree should die, and fall, and lie undisturbed on the ground, it would slowly decay. The carbon of its solid frame would pass back into the air, as gas, and the heat would escape so gradually that I could not notice it at all, unless I thrust my hand into the warm, crumbling mass.

If my tree should be cut down to-day and chopped into stove wood, it would keep a fire in my grate for many months.

Burning destroys wood substance a great deal faster than decay in the open air does, but the processes of rotting and burning are alike in this: each process releases the carbon, and gives it back to the air. It gives back also the sun's heat, stored while the tree was growing. There is left on the ground, and in the ashes on the hearth, only the mineral substance taken up in the water the roots gathered underground.

If my tree stood in swampy ground and fell over under a high wind, the water that covered it and saturated its substance would prevent decay. The carbon would not be allowed to escape as a gas to the air; the woody substance would become gradually changed into *peat.* In this form it might remain for thousands of years, and finally be changed into coal.

Whether it was burned while yet in the condition of peat, or millions of years later, when it was transformed into coal, the heat stored in its substance was liberated by the burning. The carbon and the heat went back to the air.

Every green plant we see spreads its leaves to the sun. Every stick of wood we burn, and every lump of coal, is giving back, in the form of light and heat, the energy that came from sunshine and was captured by the green leaves. How long the wood has held this store of heat we may easily compute, for we can read the age of a tree. But the age of coal we cannot accurately state. The years probably should be counted by millions, instead of thousands.

The great inland sea that covered the middle portion of the continent during the Silurian and the Devonian periods, became shallow by the deposit of vast quantities of sediment. Along the lines of the deposits of greatest thickness, a crumpling of the earth's crust lifted the first fold of the Alleghany Mountains as a great sea wall on the east, and on the western shore another formed the beginning of the Ozark Mountain system in Missouri. An island was lifted out of the sea, forming the elevated ground on which the city of Cincinnati now stands. Various other ridges and islands divided the ancient sea into much smaller bodies of water. Hemmed in by land these shallow sea-basins gradually changed into fresh-water lakes, for they no longer had connection with the ocean, and all the water they received came from rain. After centuries of freshets, and of filling in with the rock débris brought by the streams, they became great marshes, in which grew water-loving plants. Generation after generation of these plants died, and their remains, submerged by the water, were converted into peat. In the course of ages this peat became coal. This is the history of the coal measures.

There is no guesswork here. The stems of plants do not lose their microscopic structure in all the ages it has taken to transform them to coal. A thin section of coal shows under a magnifier the structure of the stems of the coal-forming plants. Moreover, the veins of coal preserve above or below them, in shales that were once deposits of mud, the branching trunks of trees, perfectly fossilized. There are no better proofs of the vegetable origin of coal than the lumps themselves. But they are plain to the naked eye, while the coal tells its story to the man with the microscope.

The fossil remains of the plants that flourished when coal was forming are gigantic, compared with plants of the same families now living. We must conclude that the climate was tropical, the air very heavy with moisture, and charged much more heavily than it is now with carbonic acid gas.

These conditions produced, in rapid succession, forests of tree ferns and horsetails and giant club mosses. These are the three types of plants out of which the coal was made. They were all rich in resin, which makes the coal burn readily. The ferns had stems as large as tree trunks. Some have been found that are eighteen inches in diameter. We know they are ferns, because the leaves are found with their fruits attached to them in the manner of present-day ferns. The stems show the well known scar by which fern leaves are joined. And the wood of these fossil fern stems is tubular in structure, just as the wood of living ferns is to-day.

Among the ferns which dominated these old marsh forests grew one kind, the scaly leaves of which covered the stems and bore their fruits on the branching tips. These giants, some of them with trunks four feet in diameter, belong to the same group of plants as our creeping club mosses, but in the ancient days they stood up among the other ferns as trees forty or fifty feet high.

The giant scouring rushes, or horsetails, had the same general characteristics as the little reed-like plants we know by those names to-day.

The highest plants of the coal period were leafy trees with nut-like fruits, that resemble the yew trees of the present. These gigantic trees were the first conifers upon the earth. They foreshadowed the pines and the other cone-bearing evergreens. Their leaves were broad and their fruits nut-like. The Japanese ginkgo, or maidenhair fern tree, is an old-fashioned conifer somewhat like those first examples of this family. Trunks sixty to seventy feet long, crowned with broad leaves and a spike of fruit, have been found lying in the upper layers of the coal-seams, and in sandstone strata that lie between the strata of coal. Peculiar circular discs, which the microscope reveals along the sides of the wood fibres of these fossil trees, prove the wood structure to be like that of modern conifers.

Generation after generation of forests lived and died in the vast spreading swamps of this era. The land sank, and freshets came here and there, drowning out all plant life, and covering the layers of peat with beds of sand or mud. When the water went down, other forests took possession, and a new coal-bed was started. It is plainly seen that flooding often put an end to coal formation. Fifteen seams of coal, one above another, is the greatest number that have been found. The veins vary from one inch to forty feet in thickness. These are separated by layers of sandstone or shale, which accumulated as sediment, covering the stumps of dead tree ferns and other growths, and preserving them as fossils to tell the story of those bygone ages as plainly as any other record could have done.

Fresh-water animals succeeded those of salt water in the swamps that formed the coal measures. Overhead, the first insects flitted among the branches of the tree ferns. Dragon-flies darted above the surface and dipped in water as they do to-day. Spiders, scorpions, and cockroaches, all air-breathing insects, were represented, but none of the higher, nectar-loving insects, like flies and bees and butterflies, were there. Flowering plants had not yet appeared on the earth. Snakelike amphibians, some fishlike, some lizard-like, and huge crocodilian forms appeared for the first time. These air-breathing swamp-dwellers could not have lived in salt water.

Fresh-water molluscs and land shells appear for the first time as fossils in the rocks of the coal measures. On the shores of the ocean, the rocks of this period show that trilobites, horseshoe crabs, and fishes still lived in vast numbers, and corals continued to form limestone. The old types of marine animals changed gradually, but the coal measures show strikingly different fossils. These rocks bear the first record of fresh-water and land animals.

THE MOST USEFUL METAL

It is fortunate for us all that, out of the half-dozen so-called useful metals, iron, which is the most useful of them all to the human race, should be also the most plentiful and the cheapest. Aluminum is abundant in the common clay and soil under our feet. But separating it is still an expensive process; so that this metal is not commercially so plentiful as iron is, nor is it cheap.

All we know of the earth's substance is based on studies of the superficial part of its crust, a mere film compared with the eight thousand miles of its diameter. Nobody knows what the core of the earth—the great globe under this surface film—is made of; but we know that it is of heavier material than the surface layer; and geologists believe that iron is an important element in the central mass of the globe.

One thing that makes this guess seem reasonable is the great abundance of iron in the earth's crust. Another thing is that meteors which fall on the earth out of the sky prove to be chiefly composed of iron. All of their other elements are ones which are found in our own rocks. If we believe that the earth itself is a fragment of the sun, thrown off in a heated condition and cooling as it flew through space, we may consider it a giant meteor, made of the substances we find in the chance meteor that strikes the earth.

Iron is found, not only in the soil, but in all plant and animal bodies that take their food from the soil. The red colour in fruits and flowers, and in the blood of the higher animals, is a form in which iron is familiar to us. It does more, perhaps, to make the world beautiful than any other mineral element known.

But long before these benefits were understood, iron was the backbone of civilization. It is so to-day. Iron, transformed by a simple process into steel, sustains the commercial supremacy of the great civilized nations of the world. The railroad train, the steel-armoured battleship, the great bridge, the towering sky-scraper, the keen-edged tool, the delicate mechanism of watches and a thousand other scientific instruments—all these things are possible to-day because iron was discovered and has been put to use.

It was probably one of the cave men, poking about in his fire among the rocks, who discovered a lump of molten metal which the heat had separated from the rest of the rocks. He examined this "clinker" after it cooled, and it interested him. It was a new discovery. It may have been he, or possibly his descendants, who learned that this metal could be pounded into other shapes, and freed by pounding from the pebbles and other impurities that clung to it when it cooled. The relics of iron-tipped spears and arrows show the skill and ingenuity of our early ancestors in making use of iron as a means of killing their prey. The earliest remains of this kind have probably been lost because the iron rusted away.

Man was pretty well along on the road to civilization before he learned where iron could be found in beds, and how it could be purified for his use. We now know that certain very minute plants, which live in quiet water, cause iron brought into that water to be precipitated, and to accumulate in the bottom of these boggy pools. In ancient days these bog deposits of iron often alternated with coal layers. Millions of years have passed since these two useful substances were laid down. To-day the coal is dug, along with the bog iron. The coal is burned to melt the iron ore and prepare it for use. It is a fortunate region that produces both coal and iron.

Bituminous coal is plentiful, and scattered all over the country, while anthracite is scarce. The cheapest iron is made in Alabama, which has its ore in rich deposits in hillsides, and coal measures close by, furnishing the raw material for coke. The result is that the region of Birmingham has become the centre of great wealth through the development of iron and coal mines.

Where water flows over limestone rock, and percolates through layers of this very common mineral, it causes the iron, gathered in these rock masses, to be deposited in pockets. All along the Appalachian Mountains the iron has been gathered in beds which are being mined. These beds of ore are usually mixed with clay and other earthy substances from which the metal can be separated only by melting. The ore is thrown into a furnace where the metal melts and trickles down, leaving behind the non-metallic impurities. It is drawn off and run into moulds, where it cools in the form of "pig" iron.

The first fuel used in the making of pig iron from the ore was charcoal. In America the early settlers had no difficulty in finding plenty of wood. Indeed, the forests were weeds that had to be cut down and burned to make room for fields of grain. The finding of iron ore always started a small industry in a colony. If there was a blacksmith, or any one else among the small company who understood working in iron, he was put in charge.

To make the charcoal, wood was cut and piled closely in a dome-shaped heap, which was tightly covered with sods, except for a small opening near the ground. In this a fire was built, and smothered, but kept going until all the wood within the oven was charred.

This fuel burned readily, with an intense heat, and without ashes. Sticks of charcoal have the form of the wood, and they are stiff enough to hold up the ore of iron so that it cannot crush out the fire. For a long time American iron was supplied by little smelters, scattered here and there. The workmen beat the melted metal on the forge, freeing it from impurities, and shaping the pure metal into useful articles. Sometimes they made it into steel, by a process learned in the Old World.

The American iron industry, which now is one of the greatest in the world, centres in Pittsburg, near which great deposits of iron and coal lie close together. The making of coke from coal has replaced the burning of charcoal for fuel. When the forests were cut away by lumbermen, the supply of charcoal threatened to give out, and experiments were made in charring coal, which resulted in the successful making of coke, a fuel made from coal by a process similar to the making of charcoal from wood. The story of the making of coke out of hard and soft coal is a long one, for it began as far back as the beginning of the nineteenth century.

In 1812 the first boat-load of anthracite coal was sent to Philadelphia from a little settlement along the Lehigh River. A mine had been opened, the owner of which believed that the black, shiny "rocks" would burn. His neighbours laughed at him, for they had tried building fires with them, and concluded that it could not be done. In Philadelphia, the owners of some coke furnaces gave the new fuel a trial, in spite of the disgust of the stokers, who thought they were putting out their fires with a pile of stones. After a little, however, the new fuel began to burn with the peculiar pale flame and intense heat that we know so well, and the stokers were convinced that here was a new fuel, with possibilities in it.

But it was hard for people to be patient with the slow starting of this hard coal. Not until 1840 did it come into general favour, following the discovery that if hot air was supplied at the draught, instead of cold, anthracite coal became a perfect fuel.

At Connellsville, Pennsylvania, a vein of coal was discovered which made coke of the very finest quality. Around this remarkable centre, coke ovens were built, and iron ore was shipped in, even from the rich beds of the Lake Superior country. But it was plain to see that Connellsville coal would become exhausted; and so experiments in coke-making from other coals were still made. When soft coal burns, a waxy tar oozes out of it, which tends to smother the fire. Early experiments with coal in melting iron ore indicated that soft coal was useless as a substitute for charcoal and coke; but later experiments proved that coke of fine quality can be made out of this bituminous soft coal, by drawing off the tar which makes the trouble. New processes were invented by which valuable gas and coal tar are taken out of bituminous coal, leaving, as a residue, coke that is equal in quality to that made from the Connellsville coal. Fortunes have been made out of the separation of the elements of the once despised soft coal. For the coke and each of its by-products, coal tar and coal gas, are commercial necessities of life.

The impurities absorbed by the melting iron ore include carbon, phosphorus, and silicon. Carbon is the chief cause of the brittleness of cast iron. The puddling furnace was invented to remove this trouble. The melted ore was stirred on a broad, basin-like hearth, with a long-handled puddling rake. The flames swept over the surface, burning the carbon liberated by the stirring. It was a hard, hot job for the man at the rake, but it produced forge iron, that could be shaped, hot or cold, on the anvil.

The next improvement was the process of pressing the hot iron between grooved rollers to rid it of slag and other foreign matters collected in the furnace. The old way was to hammer the metal free from such impurities. This was slow and hard work.

Iron was an expensive and scarce metal until the hot blast-furnace cheapened the process of smelting the ore. The puddling furnace and the grooved rollers did still more to bring it into general use. The railroads developed with the iron industry. Soon they required a metal stronger than iron. Steel was far too expensive, though it was just what was needed. Efforts were made to find a cheap way to change iron into steel. Sir Henry Bessemer solved the problem by inventing the Bessemer converter. It is a great closed retort, which is filled with melted pig iron. A draught admits air, and the carbon is all burned out. Then a definite amount of carbon, just the amount required to change iron into steel, is added, by throwing in bars of an alloy of carbon and manganese. The latter gives steel its toughness, and enables it to resist greater heat without crystallizing and thus losing its temper.

When the carbon has been put in, the retort is closed. The molten metal absorbs the alloy, and the product is Bessemer steel. In fifteen minutes pig iron can be transformed into ingot steel. The invention made possible the use of steel in the construction of bridges, high buildings, and ships. It made this age of the world the Age of Steel.

THE AGE OF REPTILES

Two big and interesting reptiles we see in the Zoo, the crocodile and its cousin, the alligator. In the everglades of Florida both are found. The crocodile of the Nile is protected by popular superstition, so it is in better luck than ours. The alligators have been killed off for their skins, and it is only a matter of time till these lumbering creatures will be found only in places where they are protected as the remnants of a vanished race. Giant reptiles of other kinds are few upon the earth now. The *boa constrictor* is the giant among snakes. The great tropical turtles represent an allied group. Most of the turtles, lizards, and snakes are small, and in no sense dominant over other creatures.

The rocks that lie among the coal measures contain fossils of huge animals that lived in fresh water and on land, the ancestors of our frogs, toads, and salamanders, a group we call amphibians. Some of these animals had the form of snakes; some were fishlike, with scaly bodies; others were lizard-like or like huge crocodiles. These were the ancestors of the reptiles, which became the rulers of land and sea during the Mesozoic Era. The rocks that overlie the coal measures contain fossils of these gigantic animals.

Strange crocodile-like reptiles, with turtle-like beaks and tusks, but no teeth, left their skeletons in the mud of the shores they frequented. And others had teeth in groups—grinders, tearers, and cutters—like mammals. These had other traits like the old-fashioned, egg-laying mammals, the duck-billed platypus, for example, that is still found in Australia. Along with the remains of these creatures are found small pouched mammals, of the kangaroo kind, in the rocks of Europe and America. These land animals saw squatty cycads, and cone-bearing trees, the ancestors of our evergreens, growing in forests, and marshes covered with luxuriant growths of tree ferns and horsetails, the fallen bodies of which formed the recent coal that is now dug in Virginia and North Carolina. Ammonites, giant sea snails, with chambered shells that reached a yard and more in diameter, and gigantic squids, swam the seas. Sea urchins, starfish, and oysters were abundant. Insects flitted through the air, but no birds appeared among the trees or beasts in the jungles. Over all forms of living creatures reptiles ruled. They were remarkable in size and numbers. There were swimming, running, and flying forms.

Banded sandstone from Calico Cañon, South Dakota

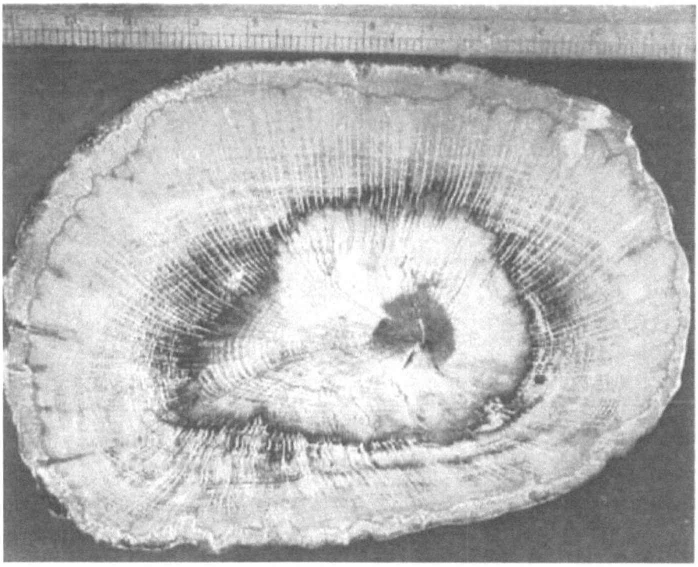

By permission of the American Museum of Natural History
Opalized wood from Utah

By permission of the American Museum of Natural History Restoration of a carnivorous Dinosaur, Allosaurus, from the Upper Jurassic and Lower Cretaceous of Wyoming. When erect the animal was about 15 feet high

The fish-reptile, *Ichthyosaurus*, was a hump-backed creature, thirty to forty feet long, with short neck, very large head, and long jaw, set with hundreds of pointed teeth. Its eye sockets were a foot across. The four short limbs were strong paddles, used for swimming. The long, slender tail ended in a flat fin. Perfect skeletons of this creature have been found. Its rival in the sea was the lizard-like *Plesiosaurus*, the small head of which was mounted on a long neck. The tail was short, but the paddles were long and powerful. No doubt this agile creature held its own, though somewhat smaller than the more massively built Ichthyosaurus.

The land reptiles called *Dinosaurs* were the largest creatures that have ever walked the earth. In the American Museum of Natural History, in New York, the mounted skeleton of the giant Dinosaur fairly takes one's breath away. It is sixty-six feet long, and correspondingly large in every part, except its head. This massive creature was remarkably short of brains.

The strangest thing about the land reptiles is the fact that certain of them walked on their hind legs, like birds, and made three-toed tracks in the mud.

Indeed, these fossil tracks, found in slate, were called bird tracks, until the bones of the reptile skeleton with the bird-like foot were discovered. Certain long grooves in the slate, hitherto unexplained, were made by the long tail that dragged in the mud.

When the mud dried, and was later covered with sediment of another kind, these prints were preserved, and when the bed of rock was discovered by quarrymen, the two kinds split apart, showing the record of the stroll of a giant along the river bank in bygone days.

The flying reptiles were still more bird-like in structure, though gigantic in size. Imagine the appearance of a great lizard with bat-like, webbed wings and bat-like, toothed jaws! The first feathered fossil bird was discovered in the limestone rock of Bavaria. It was a wonderfully preserved fossil, showing the feathers perfectly. Three fingers of each "hand" were free and clawed, so that the creature could seize its prey, and yet use its feathered wings in flight. The small head had jaws set with socketed teeth, like a reptile's, and the long, lizard tail of twenty-one bones had a pair of side feathers at each joint. This *Archeopteryx* is the reptilian ancestor of birds. During this age of the world, one branch of the reptile group established the family line of birds. The bird-like reptiles are the connecting link between the two races. How much both birds and reptiles have changed from that ancient type, their common ancestor!

I have mentioned but a few of the types of animals that make the reptilian age the wonder of all time. One after another skeletons are unearthed and new species are found. The Connecticut River Valley, with its red sandstones and shales of the Mesozoic Era, is famous among geologists, because it preserves the tracks of reptiles, insects, and crustaceans. These signs tell much of the life that existed when these flakes of stone were sandy and muddy stretches Not many bones have been found, however. The thickness of these rocks is between one and two miles. The time required to accumulate so much sediment must have been very great.

By permission of the American Museum of Natural History
Model of a three-horned Dinosaur, Triceratops, from Cretaceous of Montana. Animal in life about 25 feet long

By permission of the American Museum of Natural History
Mounting the forelegs of Brontosaurus, the aquatic Dinosaur

It is not clear just what caused the race of giant reptiles to decline and pass away. The climate did not materially change. Perhaps races grow old, and ripe for death, after living long on the earth. It seems as if their time was up; and the clumsy giants abdicated their reign, leaving dominion over the sea, the air, and the land to those animals adapted to take the places they were obliged to vacate.

THE AGE OF MAMMALS

The warm-blooded birds and mammals followed the reptiles. This does not mean that all reptiles died, after having ruled the earth for thousands of years. It means that changes in climate and other life conditions were unfavourable to the giants of the cold-blooded races, and gradually they passed away. They are represented now on the earth by lesser reptiles, which live comfortably with the wild creatures of other tribes, but which in no sense rule in the brute creation. They live rather a lurking, cautious life, and have to hide from enemies, except a few more able kinds, provided with means of defense.

There were mammals on the earth in the days of reptilian supremacy, but they were small in size and numbers, and had to avoid any open conflict with the giant reptiles, or be worsted in a fight. Now the time came when the ruling power changed hands. The mammals had their turn at ruling the lower animals. It was the beginning of things as they are to-day, for mammals still rule. But many millions of years have probably stood between the age when this group of animals first began to swarm over the earth, and the time when Man came to be ruler over all created things.

Among the reptiles of the period when the sea, the land, and the air were swarming with these great creatures were certain kinds that had traits of mammals. Others were bird-like. From these reptilian ancestors birds and mammals have sprung. No one doubts this. The fossils prove it, step by step.

Yet the rocks surprise the geologist with the suddenness with which many new kinds of mammals appeared on the earth. Possibly the rocks containing the bones of so many kinds were fortunately located. The spots may have been morasses where migrating mammals were overwhelmed while passing. Possibly conditions favored the rapid development of new kinds, and the multiplication of their numbers. Warm, moist climate furnished abundant succulent plant food for the herbivors, and these in turn furnished prey for the carnivors.

The coal formed during the Tertiary Period gives added proof that the plant life was luxuriant. The kinds of trees that grew far north of our present warm zones have left in the rocks evidence in the form of perfect leaves and cones and other fruits. For instance, magnolias grew in Greenland, and palm trees in Dakota. The temperature of Greenland was thirty degrees warmer than it is now. Our Northern States lie in a belt that must have had a climate much like that of Florida now. Europe was correspondingly mild.

A special chapter tells of the gradual development of the horse. One hundred different kinds of mammals have been found in the Eocene rocks, many of which have representative species at the same time in Europe and America. The rocks of Asia probably have similar records.

The Eocene rocks, lowest of the Tertiary strata, contain remains of animals the families of which are now extinct. Next overlying the Eocene, the Miocene rocks have fossils of animals belonging to modern families—rhinoceroses, camels, deer, dogs, cats, horses—but the genera of which are now extinct. The Pliocene strata (above the Miocene) contains fossils of animals so closely related to the wild animals now on the earth as to belong to the same genera. They differ from modern kinds only in the species, as the red squirrel is a different species from the gray.

So the record in the rocks shows a gradual approach of the mammals to the kinds we know, a gradual passing of the mighty forms that ruled by size and strength, and the coming of forms with greater intelligence, adapted to the change to a colder climate.

It sometimes happens that a farmer, digging a well on the prairie, strikes the skeleton of a monster mammal, called the *mastodon.* This very thing happened on a neighbour's farm when I was a girl, in Iowa. Everybody was excited. The owner of the land dug out every bone, careful that the whole skeleton be found. As he expected, the director of a museum was glad to pay a high price for the bones.

By permission of the American Museum of Natural History Restoration of an aquatic Dinosaur, Brontosaurus excelsus, from the Upper Jurassic and Lower Cretaceous of Wyoming. The animal in life was over 60 feet long

By permission of the American Museum of Natural History Restoration of the small carnivorous Dinosaur, Ornitholestes hermanui, catching a primitive bird Archæopteryx. Upper Jurassic and Lower Cretaceous

The mastodon was about the size of an elephant, with massive limbs, and large, heavy head that bore two stout, up-curved tusks of ivory. The creature moved in herds like the buffalo from swamp to swamp; and old age coming on, the individual, unable to keep up with the herd, sank to his death in the boggy ground. The peat accumulated over his bones, undisturbed until thousands of years elapse, and the chance digging of a well discovers his skeleton.

Frozen in the ice of northern Siberia, near the mouths of rivers, a number of mammoths have been found. These are creatures of the elephant family, and belonging to the extinct race that lived in the Quaternary Period, just succeeding the Tertiary. The ice overtook the specimens, and they have been in cold storage ever since. For this reason, both flesh and bones are preserved, a rare thing to happen, and rarer still to be seen by a scientist.

The ignorant natives made a business of watching the ice masses at the river mouth for dark spots that showed where a mammoth was encased in the ice. If an iceberg broke off near such a place, the sun might thaw the ice front of the glacier, until the hairy monster could at length be reached. His long hair served for many uses, and the wool that grew under the hair was used as a protection from the Arctic winter. The frozen flesh was eaten; the bones carved into useful tools; but the chief value of the find was in the great tusks of ivory, that curved forward and pointed over the huge shoulders. It was worth a fortune to get a pair and sell them to a buyer from St. Petersburg.

One of the finest museum specimens of the mammoth was secured by buying the tusks of the dealer, and by his aid tracing the location of the carcass, which was found still intact, except that dogs had eaten away part of one foreleg, bone and all. From this carefully preserved specimen, models have been made, exactly copying the shape and the size of the animal, its skin, hair, and other details.

The sabre-toothed tiger, the sharp tusks of which, six to eight inches long, made it a far more ferocious beast than any modern tiger of tropical jungles, was a Quaternary inhabitant of Europe and America. So was a smaller tiger, and a lion. The Irish elk, which stood eleven feet high, with antlers that spread ten feet apart at the tips, was monarch in the deer family, which had several different species on both continents. Wild horses and wild cattle, one or two of great size, roamed the woods, while rhinos and the hippopotamus kept near the water-courses. Hyenas skulked in the shadows, and acted as scavengers where the great beasts of prey had feasted. Sloths and cuirassed animals, like giant armadillos, lived in America. Among bears was one, the cave bear, larger than the grizzly. True monkeys climbed the trees. Flamingo, parrots, and tall secretary birds followed the giant *gastornis*, the ancestor of wading birds and ostriches, which stood ten feet high, but had wings as small and useless as the auk of later times.

With the entrance of the modern types of trees, came other flowering plants, and with them the insects that live on the nectar of flowers. Through a long line of primitive forms, now extinct, flowering plants and their insect friends conform to modern types. The record is written in the great stone book.

The Age of Mammals in America and Europe ended with the gradual rise of the continental areas, and a fall of temperature that ushered in the Ice Age. With the death of tropical vegetation, the giant mammals passed away.

THE HORSE AND HIS ANCESTORS

Every city has a horse market, where you may look over hundreds of animals and select one of any colour, size, or kind. The least in size and weight is the Shetland pony, which one man buys for his children to drive or ride. Another man wants a long-legged, deep-chested hunter. Another wants heavy draught-horses, with legs like great pillars under them, and thick, muscular necks—horses weighing nearly a ton apiece and able to draw the heaviest trucks. What a contrast between these slow but powerful animals and the graceful, prancing racer with legs like pipe-stems—fleet and agile, but not strong enough to draw a heavy load!

All these different breeds of horses have been developed since man succeeded in capturing the wild horse and making it help him. Man himself was still a savage, and he had to fight with wild beasts, as if he were one of them, until he discovered that he could conquer them by some power higher than physical strength. From this point on, human intelligence has been the power that rules the lower animals. Its gradual development is the story of the advance of civilization on the earth. Through unknown thousands of years it has gone on, and it is not yet finished.

By permission of the American Museum of Natural History
Restoration of a Siberian mammoth, Elephas primogenius, pursued by men of the old stone age of Europe. Late Pleistocene epoch

By permission of the American Museum of Natural History Restoration of a small four-toed ancestor of the horse family, Eohippus venticolus. Lower Eocene of Wyoming

Just when and where and how our savage ancestors succeeded in taming the wild horse of the plains and the forests of Europe or Asia is unknown. Man first made friends with the wild sheep, which were probably more docile than wild oxen and horses. We can imagine cold and hungry men seeking shelter from storms in rocky hollows, where sheep were huddled. How warm the woolly coats of these animals felt to their human fellow-creatures crowded in with them in the dark!

It is believed that the primitive men who used stone axes as implements and weapons, learned to use horses to aid them in their hunting, and in their warfare with beasts and other men. Gradually these useful animals were adapted to different uses; and at length different breeds were evolved. Climate and food supply had much to do with the size and the character of the breeds. In the Shetland Islands the animals are naturally dwarfed by the cold, bleak winters, and the scant vegetation on which they subsist. In middle Europe, where the summers are long and the winters mild, vegetation is luxurious, and the early horses developed large frames and heavy muscles. The Shetland pony and the Percheron draught-horse are the two extremes of size.

What man has done in changing the types of horses is to emphasize natural differences. The offspring of the early heavy horses became heavier than their parents. The present draught-horse was produced, after many generations, all of which gradually approached the type desired. The slender racehorses, bred for speed and endurance rather than strength, are the offspring of generations of parents that had these qualities strongly marked. Hence came the English thoroughbred and the American trotter.

We can read in books the history of breeds of horses. Our knowledge of what horses were like in prehistoric times is scant. It is written in layers of rock that are not very deep, but are uncovered only here and there, and only now and then seen by eyes that can read the story told by fossil skeletons of horses of the ages long past.

Geologists have unearthed from time to time skeletons of horses. It was Professor Marsh who spent so much time in studying the wonderful beds of fossil mammals in the western part of this country, and found among them the skeletons of many species of horses that lived here with camels and elephants and rhinoceroses and tigers, long before the time of man's coming.

How can any one know that these bones belonged to a horse's skeleton? Because some of them are like the bones of a modern horse. It is an easy matter for a student of animal anatomy to distinguish a horse from a cow by its bones. The teeth and the foot are enough. These are important and distinguishing characters. It is by peculiarities in the formation of the bones of the foot that the different species of extinct horses are recognized by geologists.

Wild horses still exist in the wilds of Russia. Remains of the same species have been dug out of the soil and found in caves in rocky regions. Deeper in the earth are found the bones of horses differing from those now living. The bones of the foot indicate a different kind of horse—an unknown species. But in the main features, the skeleton is distinctly horse-like.

In rocks of deeper strata the fossil bones of other horses are found. They differ somewhat from those found in rocks nearer the surface of the earth, and still more from those of the modern horse. The older the rocks, the more the fossil horse differs from the modern. Could you think of a more interesting adventure than to find the oldest rocks that show the skeletons of horses?

The foot of a horse is a long one, though we think of it as merely the part he walks on. A horse walks on the end of his one toe. The nail of the toe we call the "hoof." The true heel is the hock, a sharp joint like an elbow nearly half way up the leg. Along each side of the cannon, the long bone of this foot, lies a splint of bone, which is the remnant of a toe, that is gradually being obliterated from the skeleton. These two splints in the modern horse's foot tell the last chapter of an interesting story. The earliest American horse, the existence of which is proved by fossil bones, tells the first chapter. The story has been read backward by geologists. It is told by a series of skeletons, found in successive strata of rock.

The "Bad Lands" of the arid Western States are rich in fossil remains of horses. Below the surface soil lie the rocks of the Quaternary Period, which included the drift laid down by the receding glaciers and the floods that followed the melting of the ice-sheet. Under the Quaternary lie the Tertiary rocks. These comprise three series, called the Eocene, Miocene, and Pliocene, the Eocene being the oldest. In the middle region of North America, ponds and marshy tracts were filled in during the Tertiary Period, by sediment from rivers; and in these beds of clay and other rock débris the remains of fresh-water and land animals are preserved. Raised out of water, and exposed to erosive action of wind and water, these deposits are easily worn away, for they have not the solidity of older rocks. They are the crumbly Bad Lands of the West, cut through by rivers, and strangely sculptured by wind and rain. Here the fossil horses have been found.

Eohippus, the dawn horse, is the name given a skeleton found in 1880 in the lower Eocene strata in Wyoming. This specimen lay buried in a rock formation ages older than that in which the oldest known skeleton of this family had been found. Its discovery made a great sensation among scientists. This little animal, the skeleton of which is no larger than that of a fox, had four perfect toes, and a fifth splint on the forefoot, and three toes on the hind foot. The teeth are herbivorous.

Orohippus, with a larger skeleton, was found in the middle Eocene strata of Wyoming. Its feet are like those of its predecessor, except that the splint is gone. The teeth as well as the feet are more like those of the modern horse.

Mesohippus, the three-toed horse, found in the Miocene, shows the fourth toe reduced to splints, and the skeleton as big as that of a sheep. In this the horse family becomes fairly established.

Hypohippus, the three-toed forest horse, found in the middle Miocene strata of Colorado, is a related species, but not a direct ancestor of the modern horse.

Neohipparion, the three toed desert horse, from the upper Miocene strata, shows the three toes still present. But the Pliocene rocks contain fossils showing gradual reduction of the two side toes, modification of the teeth, and increase in size of the skeleton.

Protohippus and *Pliohippus*, the one-toed species from the Pliocene strata, illustrate these changes. They were about the size of small ponies.

Equus, the modern horse, was represented in the Pliocene strata by a species, now extinct, called *Equus Scotti*. This we may regard as the true wild horse of America, for it was as large as the domesticated horse, and much like it, though more like a zebra in some respects. No one can tell why these animals, once abundant in this country, became extinct at the end of the Tertiary Period. But this is undoubtedly true.

The types described form a series showing how the ancestors of the modern horse, grazing on the marshy borders of ancient ponds, lived and died, generation after generation, through a period covering thousands, possibly millions, of years. Along the sides of the crumbling buttes these ancient burying-grounds are being uncovered. Within a dozen years several expeditions, fitted out by the American Museum of Natural History, have searched the out-cropping strata in Dakota and Wyoming for bones of mammals known to have lived at the time the strata were forming in the muddy shallows along the margins of lake and marsh. Duplicate skeletons of the primitive horse types above have been found, and vast numbers of their scattered bones. Each summer geological excursions will add to the wealth of fossils of this family collected in museums.

The Tertiary rocks in Europe yield the same kind of secrets. The region of Paris overlies the estuary of an ancient river. When the strata are laid bare by the digging of foundations for buildings, bones are found in abundance. Cuvier was a famous French geologist who made extensive studies of the remains of the prehistoric animals found in this old burial-place called by scientists the Paris basin. He believed that the dead bodies floated down-stream and accumulated in the mud of the delta, where the tide checked the river's current.

Skeletons of the Hipparion, a graceful, three-toed horse, were found in numbers in the strata of the Miocene time. This animal lived in Europe while the Pliohippus and the Protohippus were flourishing in America.

A great number of species of tapir-like animals left their bones in the Paris basin, among them a three-hoofed animal which may have been the connecting link between the horse and the tapir families. Cuvier found the connecting link between tapirs and cud-chewing mammals.

THE AGE OF MAN

The hairy, woolly mammoth was one of the giant mammals that withstood the cold of the great ice flood, when the less hardy kinds were cut off by the changing climate of the northern half of Europe and America. In caves where the wild animals took refuge from their enemies, skeletons of men have been found with those of the beasts. With these chance skeletons have been found rude, chipped stone spear-heads, hammers, and other tools. With these the savage ancestors of our race defended themselves, and preyed on such animals as they could use for food. They hunted the clumsy mammoth successfully, and shared the caverns in the rocks with animals like the hyena, the sabre-toothed tiger, and the cave bear, which made these places their homes. In California a human skull was found in the bed of an ancient river, which was buried by a lava flow from craters long ago extinct. With this buried skull a few well-shaped but rough stone tools were found. This man must have lived when the great ice flood was at its height.

In southern France, caves have been opened that contained bones and implements of men who evidently lived by fishing and hunting. Bone fish-hooks showed skill in carving with the sharp edges of flint flakes. A spirited drawing of a mammoth, made on a flat, stone surface, is a proof that savage instincts were less prominent in these cave men than in those who fought the great reindeer and the mammoth farther north.

In later times men of higher intelligence formed tribes, tamed the wild horse, the ox, and the sheep, and made friends with the dog. Great heaps of shells along the shores show where the tribes assembled at certain times to feast on oysters and clams. Bones of animals used as food, and tools, are found in these heaps, called "kitchen-middens." These are especially numerous in Northern Europe. The stone implements used by these tribes were smoothly polished. A higher intelligence expressed itself also by the making of utensils out of clay. This pottery has been found in shell heaps. So the rude cave man, who was scarcely less a wild beast than the animals which competed with him for a living and a shelter from storms and cold, was succeeded by a higher man who brought the brutes into subjection by force of will and not by physical strength.

The lake-dwellers, men of the Bronze Age, built houses on piles in the lakes of Central Europe. About sixty years ago the water was low, and these relics of a vanished race were first discovered. The lake bottoms were scraped for further evidences of their life. Tools of polished stone and of bronze were taken up in considerable numbers. Stored grains and dried fruits of several kinds were found. Ornamental trinkets, weapons of hunters and warriors, and agricultural tools tell how the people lived. Their houses were probably built over the water as a means of safety from attack of beasts or hostile men.

In our country the mound-builders have left the story of their manners of life in the spacious, many-roomed tribal houses, built underground, and left with a great variety of relics to the explorers of modern times. These people worked the copper mines, and hammered and polished lumps of pure metal into implements for many uses. With these are tools of polished stone. Stores of corn were found in many mounds scattered in the Mississippi Valley.

The cliff-dwellers of the mesas of Arizona and New Mexico had habits like those of the mound-builders, and the Aztecs, a vanished race in the Southwest, at whose wealth and high civilization the invading Spaniards under Cortez marvelled. The plastered stone houses of the cliff-dwelling Indians had many stories and rooms, each built to house a tribe, not merely a family.

The Pueblo, the Moqui, and the Zuni Indians build similar dwellings to-day, isolated on the tops of almost inaccessible mesas.

Millions of years have passed since life appeared on the earth. Gradually higher forms have followed lower ones in the sea and on the land. But not all of the lower forms have gone. All grades of plants and animals still flourish, but the dominant class in each age is more highly organized than the class that ruled the preceding age.

To discover the earth's treasure, and to turn it to use; to tame wild animals and wild plants, and make them serve him; to create ever more beautiful and more useful forms in domestication; to find out the earth's life story, by reading the pages of the great stone book—these are undertakings that waited for man's coming.

PART II

THE SKY

EVERY FAMILY A "STAR CLUB"

The best family hobby we have ever had is the stars. We have a star club with no dues to pay, no officers to boss us, and only three rules:

1. We shall have nothing but "fun" in this club—no hard work. Therefore no mathematics for us!

2. We can't afford a telescope. Therefore we must be satisfied with what bright eyes can see.

3. No second-hand wonders for us! We want to see the things ourselves, instead of depending on books.

You can't imagine what pleasure we have had in one short year! The baby, of course, was too young to learn anything, and besides he was in bed long before the stars came out. But Ruth, our seven-year-old, knows ten of the fifteen brightest stars; and she can pick out twelve of the most beautiful groups or constellations. We grown-ups know all of the brightest stars, and all forty-eight of the most famous constellations. And the whole time we have given to it would not exceed ten minutes a day!

And the best part is the *way* we know the stars. The sky is no longer bewildering to us. The stars are not cold, strange, mysterious. They are friends. We know their faces just as easily as you know your playmates. For instance, we know Sirius, because he is the brightest. We know Castor and Pollux, because they are twins. We know Regulus, because he is in the handle of the Sickle. And some we know by their colours. They are just as different as President Taft, "Ty" Cobb, Horace Fletcher and Maude Adams. And quite as interesting!

What's more, none of us can ever get lost again. No matter what strange woods or city we go to, we never get "turned around." Or if we do, we quickly find the right way by means of the sun or the stars.

Then, too, our star club gives us all a little exercise when we need it most. Winter is the time when we all work hardest and have the fewest outdoor games. Winter is also the best time for young children to enjoy the stars, because it gets dark earlier in winter—by five o'clock, or long before children go to bed. It is pleasant to go out doors for half an hour before supper and learn one new star or constellation.

Again, it is always entertaining because every night you find the old friends in new places. No two nights are just the same. The changes of the moon make a great difference. Some nights you enjoy the moonlight; other nights you wish there were no moon, because it keeps you from spying out some new star. We have a little magazine that tells us all the news of the stars and the planets and the comets *before* the things happen! We pay a dollar a year for it. It is called the *Monthly Evening Sky Map.*

When we first became enthusiastic about stars, the father of our family said: "Well, I think our Star Club will last about two years. I judge it will cost us about two dollars and we shall get about twenty dollars worth of fun out of it." But in all three respects father was mistaken.

Part of the two dollars father spoke of went for a book called "The Friendly Stars," and seventy-five cents we spent for the most entertaining thing our family ever bought—a planisphere. This is a device which enables us to tell just where any star is, at any time, day or night, the whole year. It has a disc which revolves. All we have to do is to move it until the month and the day come right opposite the very hour we are looking at it, and then we can tell in a moment which stars can be seen at that time. Then we go down the street where there is a good electric light at the corner and we hold our planisphere up, almost straight overhead. The light shines through, so that we can read it, and it is just as if we had a map of the heavens. We can pick out all the interesting constellations and name them just as easily as we could find the Great Lakes or Rocky Mountains in our geography.

We became so eager not to miss any good thing that father got another book. Every birthday in our family brought a new star book, until now we have about a dozen—all of them interesting and not one of them having mathematics that children cannot understand. So I think we have spent on stars fifteen dollars more than we needed to spend (but I'm glad we did it), and I think we have had about two hundred dollars worth of fun! Yes, when I think what young people spend on ball games, fishing, tennis, skating, and all the other things that children love, I am sure our family has had about two hundred dollars worth of fun out of stars. And there is more to come!

You would laugh to know why I enjoy stars so much. I have always studied birds and flowers and trees and rocks and shells so much that I was afraid to get interested in stars. I thought it wouldn't rest me. But it's a totally different kind of science from any I ever studied! There are no families, genera, and species among the stars, thank Heaven! That's one reason they refresh me. Another is that no one can press them and put them in a herbarium, or shoot them and put them in a museum. And another thing about them that brings balm to my spirit is that no human being can destroy their beauty. No one can "sub-divide" Capella and fill it with tenements. No one can use Vega for a bill-board. Ah, well! we must not be disturbed if every member of our family has a different point of view toward the stars; we can all enjoy and love them in our own ways.

How would you like to start a Star Club like ours? You ought to be able to persuade your family to form one, because it need not cost a cent. Perhaps this book will interest them all, but the better way is for you to read about one constellation and then go out with some of the family and find it. This book does not tell about wonderful things you can never see; it tells about the wonderful things all of us can see.

I wish you success with your Star Club. Perhaps your uncles and aunts will start clubs, too. We have three Star Clubs in our family—one in New York, one in Michigan, and one in Colorado. Last winter the "Colorado Star Gazers" sent this challenge to the "New Jersey Night-Owls:" "*We bet you can't see Venus by daylight!*"

That seemed possible, because during that week the "evening star" was by far the brightest object in the sky. But father and daughter searched the sky before sunset in vain, and finally we had to ask the "Moonstruck Michiganders" how to see Venus while the sun was shining. Back came these directions on a postal-card: "Wait until it is dark and any one can see Venus. Then find some tree, or other object, which is in line with Venus and over which you can just see her. Put a stake where you stand. Next day go there half an hour before sunset, and stand a little to the west. You will see Venus as big as life. The next afternoon you can find her by four o'clock. And if you keep on you will see her day before yesterday!"

That was a great "stunt." We did it; and there are dozens like it you can do. And that reminds me that father was mistaken about our interest lasting only two years. We know that it will not die till we do. For, even if we never get a telescope, there will always be new things to see. Our club has still to catch Algol, the "demon's eye," which goes out and gleams forth every three days, because it is obscured by some dark planet we can never see. And we have never yet seen Mira the wonderful, which for some mysterious reason dies down to ninth magnitude and then blazes up to second magnitude every eleventh month.

Ah, yes, the wonders and the beauties of astronomy ever deepen and widen. Better make friends with the stars now. For when you are old there are no friends like old friends.

THE DIPPERS AND THE POLE STAR

I never heard of any boy or girl who didn't know the Big Dipper. But there is one very pleasant thing about the Dipper which children never seem to know. With the aid of these seven magnificent stars you can find all the other interesting stars and constellations. So true is this that a book has been written called "The Stars through a Dipper."

To illustrate, do you know the *Pointers*? I mean the two stars on the front side of the Dipper. They point almost directly toward the Pole star, or North star, the correct name of which is Polaris. Most children can see the Pole star at once because it is the only bright star in that part of the heavens.

But if you can't be sure you see the right one, a funny thing happens. Your friend will try to show you by pointing, but even if you look straight along his arm you can't always be sure. And then, if he tries to tell you how far one star is from another, he will try to show you by holding his arms apart. But that fails also. And so, we all soon learn the easiest and surest way to point out stars and measure distances.

The easiest way to tell any one how to find a star is to get three stars in a straight line, or else at right angles.

The surest way to tell any one how far one star is from another is by "degrees." You know what degrees are, because every circle is divided into 360 of them. And if you will think a moment, you will understand why we can see only half the sky at any one time, or 180 degrees, because the other half of the sky is on the other side of the earth. Therefore, if you draw a straight line from one horizon, clear up to the top of the sky and down to the opposite horizon, it is 180 degrees long. And, of course, it is only half that distance, or 90 degrees, from horizon to zenith. (Horizon is the point where earth and sky seem to meet, and zenith is the point straight over your head.)

Now ninety degrees is a mighty big distance in the sky. The Pole star is nothing like ninety degrees from the Dipper. It is only twenty-five degrees, or about five times the distance between the Pointers. And now comes the only thing I will ask you to remember. Look well at the two Pointers, because the distance between them, five degrees, is the most convenient "foot rule" for the sky that you will ever find. Most of the stars you will want to talk about are from two to five times that distance from some other star that you and your friends are sure of. Perhaps this is a little hard to understand. If so, read it over several times, or get some one to explain it to you, for when you grasp it, it will unlock almost as many pleasures as a key to the store you like the best.

Now, let's try our new-found ruler. Let us see if it will help us find the eighth star in the Dipper. That's a famous test of sharp eyes. I don't want to spoil your pleasure by telling you too soon where it is. Perhaps you would rather see how sharp your eyes are before reading any further. But if you can't find the eighth star, I will tell you where to look.

Look at the second star in the Dipper, counting from the end of the handle. That is a famous star called Mizar. Now look all around Mizar, and then, if you can't see a little one near it, try to measure off one degree. To do this, look at the Pointers and try to measure off about a fifth of the distance between them. Then look about one degree (or less) from Mizar, and I am sure you will see the little beauty—its name is Alcor, which means "the cavalier" or companion. The two are sometimes called "*the horse and rider*"; another name for Alcor is Saidak, which means "the test." I shall be very much disappointed if you cannot see Saidak, because it is not considered a hard test nowadays for sharp eyes.

Aren't these interesting names? Mizar, Alcor, Saidak. They sound so Arabian, and remind one of the "Arabian Nights." At first, some of them will seem hard, but you will come to love these old names. I dare say many of these star names are 4,000 years old. Shepherds and sailors were the first astronomers. The sailors had to steer by the stars, and the shepherds could lie on the ground and enjoy them without having to twist their necks. They saw and named Alcor, thousands of years before telescopes were invented, and long before there were any books to help them. They saw the demon star, too, which I have never seen. It needs patience to see those things; sharp eyes are nothing to be proud of, because they are given to us. But patience is something to be eager about, because it costs us a lot of trouble to get it.

Let's try for it. We've had a test of sight. Now let's have a test of patience. It takes more patience than sharpness of sight to trace the outline of the Little Dipper. It has seven stars, too, and the Pole star is in the end of the handle. Do you see two rather bright stars about twenty-five degrees from the Pole? I hope so, for they are the only brightish stars anywhere near Polaris. Well, those two stars are in the outer rim of the Little Dipper. Now, I think you can trace it all; but to make sure you see the real thing, I will tell you the last secret. The handle of the Big Dipper is bent *back*; the handle of the Little Dipper is bent *in.*

Now, if you have done all this faithfully, you have worked hard enough, and I will reward you with a story. Once upon a time there was a princess named Callisto, and the great god Jupiter fell in love with her. Naturally, Jupiter's wife, Juno, wasn't pleased, so she changed the princess into a bear. But before this happened, Callisto became the mother of a little boy named Arcas, who grew up to be a mighty hunter. One day he saw a bear and he was going to kill it, not knowing that the bear was really his own mother. Luckily Jupiter interfered and saved their lives. He changed Arcas into a bear and put both bears into the sky. Callisto is the Big Bear, and Arcas is the Little Bear. But Juno was angry at that, and so she went to the wife of the Ocean and said, "Please, never let these bears come to your home." So the wife of the Ocean said, "I will never let them sink beneath the waves." And that is why the Big and the Little Dipper never set. They always whirl around the Pole star. And that is why you can always see them, though some nights you would have to sit up very late.

Is that a true story? No. But, I can tell you a true one that is even more wonderful. Once upon a time, before the bear story was invented and before people had tin dippers, they used to think of the Little Dipper as a little dog. And so they gave a funny name to the Pole star. They called it Cynosura, which means "the dog's tail." We sometimes say of a great man, "he was the cynosure of all eyes," meaning that everybody looked at him. But the original cynosure was and is the Pole star, because all the stars in the sky seem to revolve around it. The two Dippers chase round it once every twenty-four hours, as you can convince yourself some night when you stay up late. So that's all for to-night.

What! You want another true story? Well, just one more. Once upon a time the Big Dipper was a perfect cross. That was about 50,000 years ago. Fifty thousand years from now the Big Dipper will look like a steamer chair. How do I know that? Because, the two stars at opposite ends of the Dipper are going in a direction different from the other five stars. How do I know that? Why, I don't know it. I just believe it. There are lots of things I don't know, and I'm not afraid to say so. I hope you will learn how to say "I don't know." It's infinitely better than guessing; it saves trouble, and people like you better, because they see you are honest. I don't know how the stars in the Big Dipper are moving, but the men who look through telescopes and study mathematics say the end stars do move in a direction opposite to the others, and they say the Dipper *must* have looked like a cross, and will look like a dipper long, long after we are dead. And I believe them.

CONSTELLATIONS YOU CAN ALWAYS SEE

There are forty-eight well known constellations, but of these only about a dozen are easy to know. I think a dozen is quite enough for children to learn. And therefore, I shall tell you how to find only the showiest and most interesting.

The best way to begin is to describe the ones that you can see almost every night in the year, because you may want to begin any month in the year, and you might be discouraged if I talked about things nobody could see in that month. There are five constellations you can nearly always see, and these are all near the Pole star.

Doubtless you think you know two of them already—the Big and the Little Dipper. Ah, I forgot to tell you that these dippers are not the real thing. They are merely parts of bigger constellations and their real names are Great Bear and Little Bear. The oldest names are the right ones. Thousands of years ago, when the Greeks named these groups of stars, they thought they looked like two bears. I can't see the resemblance.

But for that matter all the figures in the sky are disappointing. The people who named the constellations called them lions, and fishes, and horses, and hunters, and they thought they could see a dolphin, a snake, a dragon, a crow, a crab, a bull, a ram, a swan, and other things. But nowadays we cannot see those creatures. We can see the stars plainly enough, and they do make groups, but they do not look like animals. I was greatly disappointed when I was told this; but I soon got over it, because new wonders are always coming on. I think the only honest thing to do is to tell you right at the start that you cannot see these creatures very well. You will spoil your pleasure unless you take these resemblances good-naturedly and with a light heart. And you will also spoil your pleasure if you scold the ancients for naming the constellations badly. Nobody in the world would change those old names now. There is too much pleasure in them. Besides, I doubt if we could do much better. I believe those old folks were better observers than we. And I believe they had a lighter fancy.

Let us, too, be fanciful for once. I have asked my friend, Mrs. Thomas, to draw her notion of some of these famous creatures of the sky. You can draw your idea of them too, and it is pleasant to compare drawings with friends. There is only one way to see anything like a Great Bear. You have to imagine the Dipper upside down and make the handle of the Dipper serve for the Bear's tail. What a funny bear to drag a long tail on the ground! Miss Martin says he looks more like a chubby hobby-horse. You will have to make the bowl of the Dipper into hind legs and use all the other stars, somehow, to make a big, clumsy, four-legged animal. And what a monster he is! He measures twenty-five degrees from the tip of his nose to the root of his tail. Yes, all those miscellaneous faint stars you see near the Big Dipper belong to the Great Bear.

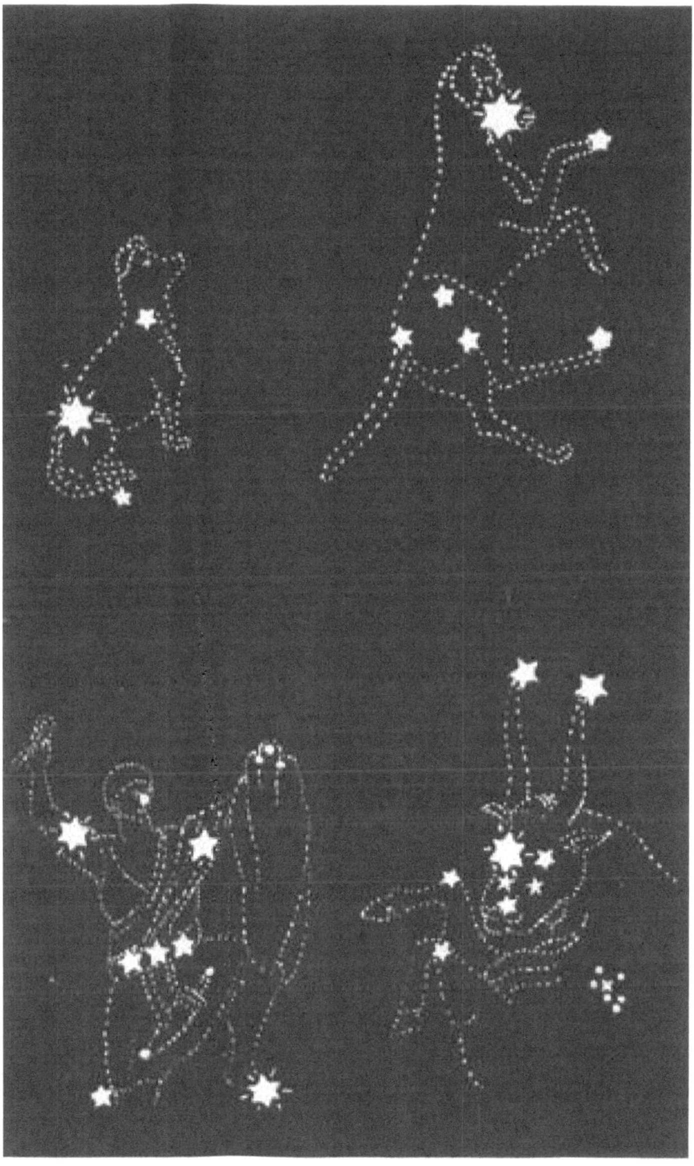

Orion fighting the Bull. Above are Orion's two dogs

The Little Bear, the Queen in her chair, the Twins and the Archer

How the Great Bear looked to the people who named it thousands of years ago, we probably shall never know. They left no books or drawings, so far as I know. But in every dictionary and book on astronomy you can find these bears and other animals drawn so carefully and beautifully that it seems as if they *must* be in the sky, and we must be too dull to see them. It is not so. Look at the pictures in this book and, you will see that the stars do not outline the animals. Many of them come at the wrong places. And so it is with all the costly books and charts and planispheres. It is all very interesting, but it isn't true. It's just fancy. And when you once understand that it isn't true, you will begin to enjoy the fancy. Many a smile you will have, and sometimes a good laugh. For instance, the English children call the Dipper "Charles's wain" or "the wagon." And the Romans called it "the plough." They thought of those seven stars as oxen drawing a plough.

Well, that's enough about the two Bears. I want to tell you about the other three constellations you can nearly always see. These are the Chair, the Charioteer, and Perseus (pronounced *per'soos*).

The Chair is the easiest to find, because it is like a very bad W, and it is always directly opposite the big Dipper, with the Pole star half way between the two constellations. There are five stars in the W, and to make the W into a Chair you must add a fainter star which helps to make the square bottom of the chair. But what a crazy piece of furniture! I have seen several ways of drawing it, but none of them makes a comfortable chair. I should either fall over backward, or else the back of the chair would prod me in the small of my back. The correct name of this constellation is Cassiopeia's Chair.

I think this is enough to see and enjoy in one night. To-morrow night let us look for the Charioteer.

I love the Charioteer for several reasons. One is that it makes a beautiful pentagon, or five-sided figure, with its five brightest stars. Another is that it contains the second-brightest star in the northern part of the heavens, Capella. The only star in the north that is brighter is Vega, but Vega is bluish white or creamy.

If you haven't already found the five-sided figure, I will tell you how to find Capella. Suppose you had a gun that would shoot anything as far as you wish. Shoot a white string right through the Pointers and hit the Pole star. Then place your gun at the Pole star and turn it till it is at right angles to that string you shot. Aim away from the big Dipper, shoot a bullet forty-five degrees and it will hit Capella.

If that plan doesn't work, try this. Start with the star that is at the bottom of the Dipper and nearest the handle. Draw a line half-way between the two

Pointers and keep on till you come to the first bright star. This is Capella, and the distance is about fifty degrees.

Capella means "a kid," or "little goat," and that reminds me of the third reason why I enjoy so much the constellation of which Capella is the brightest star. In the old times they sometimes called this five-sided figure "the goat-carrier." And the shepherds thought they could see a man carrying a little goat in his left hand. I am sure you can see the kid they meant. It is a triangle of faint stars which you see near Capella. That's enough for to-night.

To-morrow night let us look for Perseus. I dare say you know that old tale about Perseus rescuing the princess who was chained upon a rock. (He cut off the snaky head of Medusa and showed it to the dragon that was going to devour the princess, and it turned the monster to stone. Remember?) Well, there are constellations named after all the people in that story, but although they contain many showy stars, I could never make them look like a hero, a princess, a king, and a queen. I do not even try to trace out all of Perseus. For I am satisfied to enjoy a very beautiful part of it which is called the "Arc of Perseus."

An arc, you know, is a portion of a circle. And the way to find this arc is to draw a curve from Capella to Cassiopeia. On nights that are not very clear I can see about seven stars in this Arc of Perseus. And the reason I love it so much is that it is the most beautiful thing, when seen through an opera-glass, that I know. You could never imagine that a mere opera-glass would make such a difference. The moment I put it to my eyes about a dozen more stars suddenly leap into my sight in and near the Arc of Perseus. That's enough. No more stories to-night.

WINTER CONSTELLATIONS

By winter constellations I mean those you can see in winter at the pleasantest time—the early evening. And I want you to begin with the Northern Cross. I hope you can see this before Christmas, for, after that, it may be hidden by trees or buildings in the west and you may not see it again for a long while. This is because the stars seem to rise in the east and set in the west. To prove this, choose some brilliant star you can see at five or six o'clock; get it in line with some bush or other object over which you can just see it. Put a stake where you stand, and then go to the same spot about eight o'clock or just before you go to bed. You can tell at once how much the star seems to have moved westward.

Another thing, every star rises four minutes later every night, and therefore the sky looks a little different at the same hour every evening. The stars in the north set for a short time only, but when those toward the south set they are gone a long time. For instance, the brightest star of all is Sirius, the Dog Star, which really belongs to the southern hemisphere. There are only about three months in the year when children who go to bed by seven o'clock can see it—January, February, and March.

So now you understand why I am so eager that you should not miss the pleasure of seeing the famous Northern Cross. But although it is a big cross, and easy to find, after you know it, I have never yet known a boy who could show it to another boy simply by pointing at it. The surest and best way to find it is learn three bright stars first—Altair, Vega, and Deneb.

Altair is the brightest star in the Milky Way. It is just at the edge of the Milky Way, and you are to look for three stars in a straight line, with the middle one brightest. Those three stars make the constellation called "the Eagle." The body of the eagle is Altair, and the other two stars are the wings. I should say that Altair is about five degrees from each of his companions. It is worth half an hour's patient search to find the Eagle. Now these three stars in the Eagle point straight toward the brightest star in the northern part of the sky—Vega.

To make sure of it, notice four fainter stars near it which make a parallelogram—a sort of diamond. These stars are all part of a constellation called "the Lyre." If you try to trace out the old musical instrument, you will be disappointed; but here is a game worth while. Can you see a small triangle made by three stars, of which Vega is one? Well, one of those stars is double, and with an opera-glass you can see which it is. On very clear nights some people with very sharp eyes can see them lying close together, but I never could.

At last we are ready to find the celebrated Northern Cross. First draw a line from Altair to Vega. Then draw a line at right angles to this, until you come to another bright star—Deneb—which is about as far from Vega as Vega is from Altair. Now this beautiful star, Deneb, is the top of the Northern Cross. I can't tell you whether the Cross will be right or wrong side up when you see it, or on its side. For every constellation is likely to change its position during the night, as you know from watching the Dipper. But you can tell the Cross by these things. There are six stars in it. It is like a kite made of two sticks. There are three stars in the crosspiece and four in the long piece. Deneb, the brightest star in the cross, is at the top of the long stick.

But you mustn't expect to see a perfect cross. There is one star that is a little out of place, and sometimes my fingers fairly "itch to put it where it belongs." It is the one that ought to be where the long stick of your kite is tacked to the crosspiece. And one of the stars is provokingly faint, but you can see it. Counting straight down the long piece, it is the third one from Deneb that is faint. It is where it ought to be, but I should like to make it brighter. Have you the Cross now? If not, have patience. You can't be a "true sport" unless you are patient. You can't be a great ball-player, or hunter, or any thing else, without resisting, every day, that sudden impulse to "quit the game" when you lose. Be a "good loser," smile and try again. That is better than to give up, or to win by cheating or sharp practice.

This is the last thing I want you to see in the northern part of the sky; and if you have done a good job, let us celebrate by having a story.

Once upon a time a cross didn't mean so much to the world as it does now. That was before Christ was born. In those old times people did not think of the Northern Cross as a cross. They thought of it as a Swan, and you can see the Swan if you turn the Cross upside down. Deneb will then be in the tail of the Swan, and the two stars which used to be at the tips of the crosspiece now become the wings. Is that a true story? Yes. If we lived in Arabia the children there could tell us what Deneb means. It means "the tail."

Another story? Well, do you see the star in the beak of the Swan, or foot of the Cross? What color is it? White? Well, they say this white star is really made up of two stars—one yellow and the other blue. That is one reason I want to buy a telescope when I can afford it, for even the smallest telescope will show that. And Mr. Serviss says that even a strong field-glass will help any one see this wonder.

I can't tell you about all the winter constellations in one chapter. We have made friends of the northern ones. Now let's see the famous southern ones. And let's start a new chapter.

ORION, HIS DOGS, AND THE BULL

The most gorgeous constellation in the whole sky is Orion. I really pity any one who does not know it, because it has more bright stars in it than any other group. Besides, it doesn't take much imagination to see this mighty hunter fighting the great Bull. I dare say half the people in the United States know Orion and can tell him as quick as they see him by the famous "belt of Orion."

This belt is made of three stars, each of which is just one degree from the next. That is why the English people call these three stars "the ell and yard." Another name for them is "the three kings." You can see the "sword of Orion" hanging down from his belt.

As soon as you see these things you will see the four bright stars that outline the figure of the great hunter, but only two of them are of the first magnitude. The red one has a hard name—Betelgeuse (pronounced *bet-el-guz´*). That is a Frenchified word from the Arabic, meaning "armpit," because this star marks the right shoulder of Orion. The other first-magnitude star is the big white one in the left foot. Its name is Rigel (pronounced *re´-jel*) from an Arabian word meaning "the foot."

You can see the giant now, I am sure. Over his left arm hangs a lion's skin which he holds out to shield him from the bull's horns. See the shield—about four rather faint stars in a pretty good curve? Now look for his club which he holds up with his right hand so as to smite the bull. See the arm and the club—about seven stars in a rather poor curve—beyond the red star Betelgeuse? Now you have him, and isn't he a wonder!

It is even easier to see the Bull which is trying to gore Orion. Look where Orion is threatening to strike, and you will see a V. How many stars in that V? Five. And which is the brightest? That red one at the top of the left branch of the V? Yes. That V is the face of the Bull and that red star is the baleful eye of the angry Bull which is lowering his head and trying to toss Orion. The name of that red eye is Aldebaran (pronounced *al-deb´-ar-an*).

I wish Aldebaran meant "red eye," but it doesn't. It is an old Arabian word meaning the "hindmost," or the "follower," because every evening it comes into view about an hour after you can see the famous group of stars called the Pleiades, which are in the shoulder of the Bull.

I do not care to trace the outline of this enormous bull, but his horns are a great deal longer than you think at first. If you will extend the two arms of that V a long way you will see two stars which may be called the tips of his horns. One of these stars really belongs in another constellation—our old friend the Charioteer, the one including Capella. Wow! what a pair of horns!

But now we come to the daintiest of all constellations—the Seven Sisters, or Pleiades (pronounced *plee´-a-deez*).

I can see only six of them, and there is a famous old tale about the "lost Pleiad." But I needn't describe them. Every child finds them by instinct. Some compare them to a swarm of bees; some to a rosette of diamonds; some to dewdrops. But I would not compare them to a dipper as some do, because the real Little Dipper is very different. The light that seems to drip from the Pleiades is quivering, misty, romantic, magical. No wonder many children love the Pleiades best of all the constellations. No wonder the poets have praised them for thousands of years. The oldest piece of poetry about them that I know of was written about 1,500 years before Christ. You can find it in the book of Job. But the most poetic description of the Pleiades that I have ever read is in Tennyson's poem "Locksley Hall," in which he says they "glitter like a swarm of fireflies tangled in a silver braid."

There are a great many old tales about the lost Pleiad. One is that she veiled her face because the ancient city of Troy was burned. Another story says she ceased to be a goddess when she married a man and became mortal. Some people think she was struck by lightning. Others believe the big star, Canopus, came by and ran away with her. Still others declare she was a new star that appeared suddenly once upon a time, and after a while faded away.

For myself, I do not believe any of these stories. One reason why I don't is that a seventh star is really there, and many people can really see it. Indeed, there are some people so sharp-eyed that on clear nights they can see anywhere from eight to eleven. And, what is more, they can draw a map or chart showing just where each star seems to them to be.

But the most wonderful stories about the Pleiades are the true stories. One is that there are really more than 3,000 stars among the Pleiades. Some of them can be seen only with the biggest telescopes. Others are revealed only by the spectroscope. And some can be found only by means of photography.

But the most amazing thing about the Pleiades is the distances between them. They look so close together that you would probably say "the moon seems bigger than all of them put together." Sometimes the moon comes near the Pleiades, and you expect that the moon will blot them all out. But the astronomers say the full moon sails through the Pleiades and covers only one of them at a time, as a rule. They even say it is possible for the moon to pass through the Pleiades without touching one of them! I should like to see that. If anything like it is going to occur, the magazine I spoke of in the first chapter will tell me about it. And you'd better believe I will stay up to see that, if it takes all night!

There are two more constellations in the southern part of the sky that ought to be interesting, because they are the two hunting dogs that help Orion fight the Bull. But I can't trace these animals, and I don't believe it is worth while. The brightest stars in them everybody can see and admire—Sirius, the Bigger Dog, and Procyon, the Smaller Dog.

Every one ought to know Sirius, because he is the brightest star of all. (Of course, he is not so bright as Venus and Jupiter, but they are planets.) To find him, draw a line from the eye of the Bull through the belt of Orion and extend it toward the southeast about twenty degrees. They call him the Dog star because he follows the heels of Orion. And people still call the hottest days of summer "dog days" because 400 years before Christ the Romans noticed that the Dog star rose just before the sun at that time. The Romans thought he chased the sun across the sky all day and therefore was responsible for the great heat. But that was a foolish explanation. And so is the old notion that dogs are likely to go mad during the dog days "because the dog star is in the ascendant." So is the idea that Sirius is an unlucky star.

There are no lucky or unlucky stars. These are all superstitions, and we ought to be ashamed to believe any superstition. Yet for thousands of years before we had public schools and learned to know better, people believed that every one was born under a lucky star or an unlucky one, and they believe that farmers ought to plant or not plant, according to the size of the moon. Now we know that is all bosh. Those old superstitions have done more harm than good. One of the most harmful was the belief in witches. Let us resolve never to be afraid of these old tales, but laugh at them.

Why should anybody be afraid of anything so lovely as Sirius? I used to think Sirius twinkled more than any other star. But that was bad reasoning on my part. I might have noticed that every star twinkles more near the horizon than toward the zenith. I might have noticed that stars twinkle more on clear, frosty nights than when there is a little uniform haze. And putting those two facts together I might have reasoned that the stars never really twinkle at all; they only seem to. I might have concluded that the twinkling is all due to the atmosphere—that blanket of air which wraps the earth around. The nearer the earth, the thicker the air, and the more it interferes with the light that comes to us from the stars.

They say that Sirius never looks exactly alike on two successive nights. "It has a hundred moods," says Mr. Serviss, "according to the state of the atmosphere. By turns it flames, it sparkles, it glows, it blazes, it flares, it flashes, it contracts to a point, and sometimes when the air is still, it burns with a steady white light." (Quotation somewhat altered and condensed.)

It is a pity that so fine a star as Procyon should be called the "Smaller Dog," because it suffers unjustly by comparison with Sirius. If it were in some other part of the sky we might appreciate it more, because it is brighter than most of the fifteen first-magnitude stars we can see. My brother William has grown to love it, but perhaps that is because he always "sympathizes with the under dog." He was the youngest brother and knows. And curiously enough he was nicknamed "the dog"—just why, I don't know.

To find Procyon, drawn a line from Sirius northeast about twenty degrees. And to make sure, draw one east from Betelgeuse about the same distance. These three stars make a triangle of which the sides are almost equal.

The name Procyon means "before the dog" referring to the fact that you can see him fifteen or twenty minutes earlier every night than you can see Sirius.

The only kind word about Procyon I have heard in recent years was in connection with that miserable business of Dr. Cook and the North Pole. A Captain Somebody-or-other was making observations for Dr. Cook, and he wanted to know what time it was. He had no watch and didn't want to disturb any one. So he looked out of the window and saw by the star Procyon that it was eleven o'clock.

That sounds mysterious, but it is easy if you have a planisphere like ours. Last winter when we were all enjoying Orion, the Bull, and the two Dogs, I used to whirl the planisphere around to see where they would be at six o'clock at night, at eight, at ten, at midnight, and even at six o'clock in the morning. And so, if I waked up in the night I could tell what time it was without even turning my head. Sometimes I looked out of my window, saw Orion nearly overhead and knew it must be midnight. And sometimes I woke up just before daybreak and saw the great Bull backing down out of sight in the west, the mighty Hunter still brandishing his club, and his faithful Dogs following at his heels.

SEVEN FAMOUS CONSTELLATIONS

There are only seven more constellations that seem to me interesting enough for every one to know and love all his life. These are:

The Lion (Spring)	The Herdsman (Summer)
The Twins (Spring)	The Northern Crown (Summer)
The Virgin (Summer)	The Scorpion (Summer)

Southern Fish (Autumn)

I have named the seasons when, according to some people, these constellations are most enjoyable. But these are not the only times when you can see them. (If you had that seventy-five-cent planisphere, now, you could always tell which constellations are visible and just where to find them.) No matter what time of year you read this chapter, it is worth while to go out and look for these marvels. You can't possibly miss them all.

Have you ever seen a Sickle in the sky? It's a beauty, and whenever I have seen it it has been turned very conveniently for me, because I am left-handed. It is so easy to find that I am almost ashamed to tell. But if you need help, draw a line through the Pointers backward, away from the Pole star, about forty degrees, and it will come a little west of the Sickle. The Sickle is only part of the Lion—the head and the forequarters. Only fanciful map-makers can trace the rest of the Lion. The bright star at the end of the handle is Regulus, which means "king," from the stupid old notion that this star ruled the lives of men. To this day people speak of the "Royal Star," meaning Regulus. And at the end of this chapter I will tell you about three other stars which the Persians called "royal stars."

Another constellation which children particularly love is the Twins—Castor and Pollux. But the sailors got there first! For thousands of years the twins have been supposed to bring good luck to sailors. I don't believe a word of it. But I do know that sailors gloat over Castor and Pollux, and like them better than any other stars. The whole constellation includes all the stars east of the Bull and between the Charioteer and Procyon. But another way to outline the twins is to look northeast of Orion where you will see two rows of stars that run nearly parallel. To me the brothers seem to be standing, but all the old picture-makers show them sitting with their arms around each other, the two brightest stars being their eyes. The eyes are about five degrees apart—the same as the Pointers.

Pollux is now brighter than Castor, but for thousands of years it was just the other way. It is only within three hundred years that this change has taken place. Whether Castor has faded or Pollux brightened, or both, I do not know. Anyhow, Castor is not quite bright enough to be a first magnitude star. Three hundred years is a short time in the history of man, and only a speck in the history of the stars. Three hundred years ago they killed people in Europe just because of the church they went to. That was why the Pilgrim Fathers sailed from England in 1620, and made the first permanent settlement in America, except, of course, Jamestown, Va., in 1607.

There are plenty of stories about old Castor and Pollux, and, like all the other myths, they conflict, more or less. But all agree that these two brothers went with Jason in the ship Argo, shared his adventures and helped him get the golden fleece. And all agree that Castor and Pollux were "born fighters." And that is why the Roman soldiers looked up to these stars and prayed to them to help them win their battles.

Now for the four summer constellations every one ought to know. The first thing to look for is two famous red or reddish stars—Arcturus and Antares.

The way you find Arcturus is amusing. Look for the Big Dipper and find the star at the bottom of the dipper nearest the handle. Got it? Now draw a curve that will connect it with all the stars in the handle, and when you come to the end of the handle keep on till you come to the first very bright star—about twenty-five degrees. That is the monstrous star Arcturus, probably the biggest and swiftest star we can ever see with the naked eye in the northern hemisphere. He is several times as big as our sun, and his diameter is supposed to be several million miles. He is called a "runaway sun," because he is rushing through space at the rate of between two hundred and three hundred miles a second. That means between seventeen and thirty-four million miles a day!

He is coming toward us, too! At such a rate you might think that Arcturus would have smashed the earth to pieces long ago. But he is still very far away, and there is no danger. Some people say that if Job were to come to life, the sky would seem just the same to him as it did 3,400 years ago. The only difference he might notice would be in Arcturus. That would seem to him out of place by a distance about three times the apparent diameter of the moon.

Some people believe this because Job said, "Canst thou guide Arcturus with his sons?" and therefore they imagine that he meant this red star. But I believe he meant the Big Dipper. For in King James's time, when the Bible was translated into English, the word "Arcturus" meant the Big Dipper or rather the Great Bear. And for centuries before it meant the Great Bear. One proof of it is that "Arcturus" comes from an old Greek word meaning "bear"—the same word from which we get arctic. It is only within a few hundred years that astronomers have agreed to call the Great Bear "Ursa Major," and this red star Arcturus. So I think all the books which say Job mentioned this red star are mistaken. I believe Webster's Dictionary is correct in this matter, and I believe the Revised Version translates Job's Hebrew phrase more correctly when it says, "Canst thou guide the Bear with her train?"

Anyhow, Arcturus is a splendid star—the brightest in the constellation called the "Herdsman" or Boötes. It is not worth while to trace the Herdsman, but here is an interesting question. Is Arcturus really red? The books mostly say he is yellow. They say he looks red when he is low in the sky, and yellow when he is high. How does he look to you? More yellow than red?

Well, there's no doubt about Antares being red. To find him, draw a long line from Regulus through Arcturus to Antares, Arcturus being more than half way between the other two. But if Regulus and the Sickle are not visible, draw a line from Altair, at right angles to the Eagle, until you come to a bright star about sixty degrees away. You can't miss Antares, for he is the only red star in that part of the sky.

Antares belongs to a showy constellation called the Scorpion. I cannot trace all the outline of a spider-like creature, but his poisonous tail or "stinger" is made by a curved line of stars south and east of Antares. And you can make a pretty fan by joining Antares to several stars in a curve which are west of Antares and a little north. There is an old tale that this Scorpion is the one that stung Orion to death when he began to "show off" and boast that there was no animal in the world that could kill him.

Another very bright star in the southern part of the sky is Spica. To find it, start with the handle of the Dipper, and making the same backward curve which helped you to find Arcturus, keep on till you come to the white star Spica—say thirty degrees beyond Arcturus. This is the brightest star in the constellation called "the Virgin." It is not worth while trying to trace her among nearly forty faint stars in this neighbourhood. But she is supposed to be a winged goddess who holds up in her right hand an *ear of wheat*, and that is what Spica means.

Now for an autumn constellation—the Southern Fish. I don't care if you fail to outline a fish, but I do want you to see the bright star that is supposed to be in the fish's mouth. And I don't want you to balk at its hard name—Fomalhaut (pronounced *fo´-mal-o*). It is worth a lot of trouble to know it as a friend. To find it, you have to draw an exceedingly long line two-thirds of the way across the whole sky. Start with the Pointers. Draw a line through them and the Pole star and keep clear on until you come to a solitary bright star rather low down in the south. That is Fomalhaut. It looks lonely and is lonely, even when you look at it through a telescope.

And now for the last story. Once upon a time the Persians thought there must be four stars that rule the lives of men. So they picked out one in the north and one in the south and one in the east and one in the west, just as if they were looking for four bright stars to mark the points of the compass. If you want to find them yourself without my help don't read the next sentence, but shut this book and go out and see. Then write down on a piece of paper the stars you have selected and compare them with the list I am about to give. Here are the four royal stars of the Persians: Fomalhaut for the north, Regulus for the south, Aldebaran for the east, and Antares for the west.

Why doesn't this list agree with yours? Because Persia is so far south of where we live. Ah, there are very few things that are absolutely true. Let's remember that and not be too sure: for everything depends upon the point of view! I hope you will see Fomalhaut before Christmas, before he disappears in the west. He is with us only five months and is always low—near the horizon. But the other seven months in the year he gladdens the children of South America and the rest of the southern hemisphere, for they see him sweeping high and lonely far up into their sky and down again.

But the loveliest of all the constellations described in this chapter is the Northern Crown. It is not a perfect crown—only about half a circle—but enough to suggest a complete ring. Look for it east of Arcturus. I can see seven or eight stars in the half-circle, one of which is brighter than all the others. That one is called "the Pearl." The whole constellation is only fifteen degrees long, but "fine things come in small packages"; and children grow to love the Northern Crown almost as much as they love the Pleiades.

THE TWENTY BRIGHTEST STARS

If you have seen everything I have described so far, you have reason to be happy. For now you know sixteen of the most famous constellations and fifteen of the twenty brightest stars. There are only twenty stars of the first magnitude. "Magnitude" ought to mean size, but it doesn't. It means brightness—or rather the apparent brightness—of the stars when seen by us. The word magnitude was used in the old days before telescopes, when people thought the brighter a star is the bigger it must be. Now we know that the nearer a star is to us the brighter it is, and the farther away the fainter. Some of the bright stars are comparatively near us, some are very far. Deneb and Canopus are so far away that it takes over three hundred years for their light to reach us. What whoppers they must be—many times as big as our sun.

Here is a full list of the twenty stars of the first magnitude arranged in the order of their brightness. You will find this table very useful.

Stars	Pronounced	Constellation	Interesting facts
Sirius	*sir´i-us*	Big Dog	Brightest star. Nearest star visible in Northern hemisphere
Canopus*	*ca-no´pus*	Ship Argo	Perhaps the largest body in universe
Alpha Centauri*	*al´fa sen-taw´re*	Centaur	Nearest star. Light four years away
Vega	*ve´ga*	Lyre	Brightest star in the Northern sky. Bluish
Capella	*ca-pell´a*	Charioteer	Rivals Vega, but opposite the pole Yellowish
Arcturus	*ark-tu´rus*	Herdsman	Swiftest of the bright stars. 200 miles a second
Rigel	*re´jel*	Orion	Brightest star in Orion. White star in left foot
Procyon	*pro´si-on*	Little Dog	Before the dog. Rises a little before Sirius
Achernar*	*a-ker´nar*	River Po	Means the end of the river

Beta Centauri*	*ba´ta sen-taw´re*	Centaur	This and its mate point to the Southern Cross
Altair	*al-tare´*	Eagle	Helps you find Vega and Northern Cross
Betelgeuse	*bet-el-guz´*	Orion	Means "armpit." The red star in the right shoulder
Alpha Crucis*	*al´fa cru´sis*	Southern Cross	At the base of the most famous Southern constellation
Aldebaran	*al-deb´a-ran*	Bull	The red eye in the V
Pollux	*pol´lux*	Twins	Brighter than Castor
Spica	*spi´ca*	Virgin	Means ear of wheat
Antares	*an-ta´rez*	Scorpion	Red star. Name means "looks like Mars"
Fomalhaut	*fo´mal-o*	Southern Fish	The lonely star in the Southern sky
Deneb	*den´eb*	Swan	Top of Northern Cross, or tail of Swan
Regulus	*reg´u-lus*	Lion	The end of the handle of the Sickle

The five stars marked * belong to the Southern hemisphere, and we can never see them unless we travel far south. Last winter I went to Florida and saw Canopus, but to see the Southern Cross you should cross the Tropic of Cancer.

HOW TO LEARN MORE

All I can hope to do in this book is to get you enthusiastic about astronomy. I don't mean "gushy." Look in the dictionary and you will find that the enthusiast is not the faddist. He is the one who sticks to a subject for a lifetime.

Nor do I care a rap whether you become an astronomer—or even buy a telescope. There will be always astronomers coming on, but there are too few people who know and love even a few of the stars. I want you to make popular astronomy a life-long hobby. Perhaps you may have to drop it for ten or fifteen years. Never mind, you will take up the study again. I can't expect you to read a book on stars if you are fighting to make a living or support a family, unless it really rests you to read about the stars. It does rest me. When things go wrong at the office or at home, I can generally find rest and comfort from music. And if the sky is clear, I can look at the stars, and my cares suddenly seem small and drop away.

Let me tell you why and how you can get the very best the stars have to teach you, without mathematics or telescope. Follow this programme and you need never be afraid of hard work, or of exhausting the pleasures of the subject. Go to your public library and get one of the books I recommend in this chapter, and read whatever interests you. I don't care whether you take up planets before comets or comets before planets, but whatever you do do it well. Soak the interesting facts right in. Nail them down. See everything the book talks about. Make notes of things to watch for. Get a little blank book and write down the date you first saw each thing of interest. Write down the names of the constellations you love most. Before you lay down any star book you are reading, jot down the most wonderful and inspiring thing you have read—even if you have only time to write a single word that may recall it all to you. Treasure that little note book as long as you live. Every year it will get more precious to you.

Now for the books:

1. *Martin. The Friendly Stars.* Harper & Brothers, New York, 1907.

This book teaches you first the twenty brightest stars and then the constellations. I cannot say that this, or any other, is the "best book," but it has helped me most, and I suppose it is only natural that we should love best the first book that introduces us to a delightful subject.

2. *Serviss. Astronomy with the Naked Eye.* Harper & Brothers, New York, 1908.

This teaches you the constellations first and the brightest stars incidentally. Also it gives the old myths.

3. *Serviss. Astronomy with an Opera-Glass.* D. Appleton & Co., New York, 1906.

4. *Serviss. Pleasures of the Telescope.* D. Appleton & Co., New York, 1905.

5. *Milham. How to Identify the Stars.* The Macmillan Co., New York, 1909.

This gives a list of eighty-eight constellations, including thirty-six southern ones, and has tracings of twenty-eight.

6. *Elson. Star Gazer's Handbook.* Sturgis & Walton Co., New York, 1909.

About the briefest and cheapest. Has good charts and makes a specialty of the myths.

7. *Serviss. Curiosities of the Sky.* Harper & Brothers, New York.

Tells about comets, asteroids, shooting stars, life on Mars, nebulæ, temporary stars, coal-sacks, Milky Way, and other wonders.

8. *Ball. Starland.* Ginn & Co., Boston, New York, etc., 1899.

This tells about a great many interesting experiments in astronomy that children can make.

If I had only a dollar or less to spend on astronomy I should buy a planisphere. I got mine from Thomas Whittaker, No. 2 Bible House, New York. It cost seventy-five cents, and will tell you where to find any star at any time in the year. It does not show the planets, however. A planisphere that will show the planets costs about five dollars. However, there are only two very showy planets, viz., Venus and Jupiter. Any almanac will tell you (for nothing) when each of these is morning star, and when each of them is evening star.

The best newspaper about stars, as far as I know, is a magazine called *The Monthly Evening Sky Map*, published by Leon Barritt, 150 Nassau St., New York. It costs a dollar a year. It gives a chart every month, showing all the planets, and all the constellations. Also it tells you about the interesting things, like comets, before they come.

Good-bye. I hope you will never cease to learn about and love the earth and the sky. Perhaps you think you have learned a great deal already. But your pleasures have only begun. Wait till you learn about how the world began, the sun and all his planets, the distances between the stars, and the millions of blazing suns amid the Milky Way!

THE END

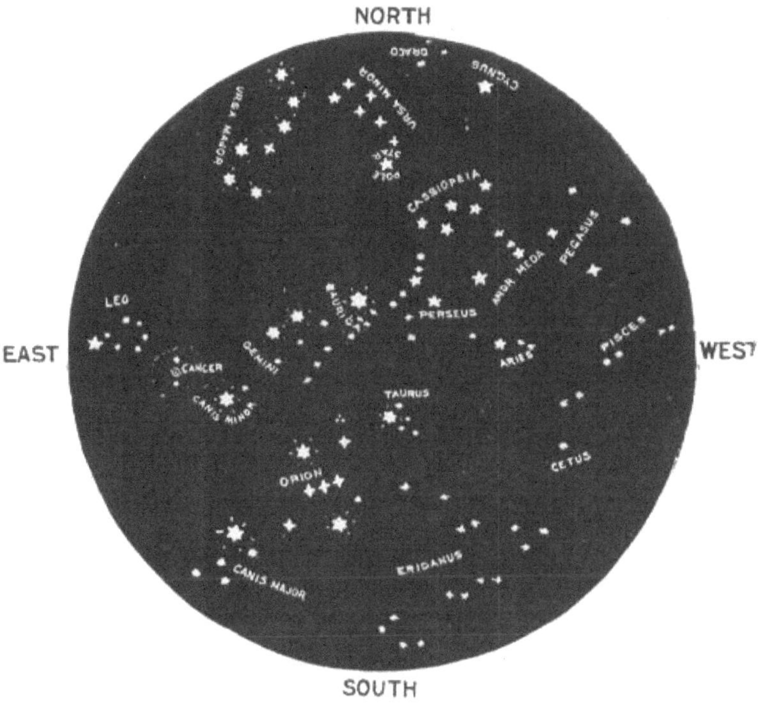

THE SKY IN WINTER

NOTE.—These simplified star maps are not as accurate as a planisphere, but they may be easier for children. All star maps are like ordinary maps, except that east and west are transposed. The reason for this is that you can hold a star map over your head, with the pole star toward the north, and the map will then match the sky. These maps contain some constellations that are only for grown-ups to study. The Winter constellations every child should know are:

AURIGA, the Charioteer
CANIS MAJOR, the Big Dog
CANIS MINOR, the Little Dog
CASSIOPEIA, the Queen in Her Chair
CYGNUS, the Swan
LEO, the Lion
ORION, the Hunter
PERSEUS, Which Has the Arc
TAURUS, the Bull
URSA MAJOR, the Great Bear
URSA MINOR, the Little Bear

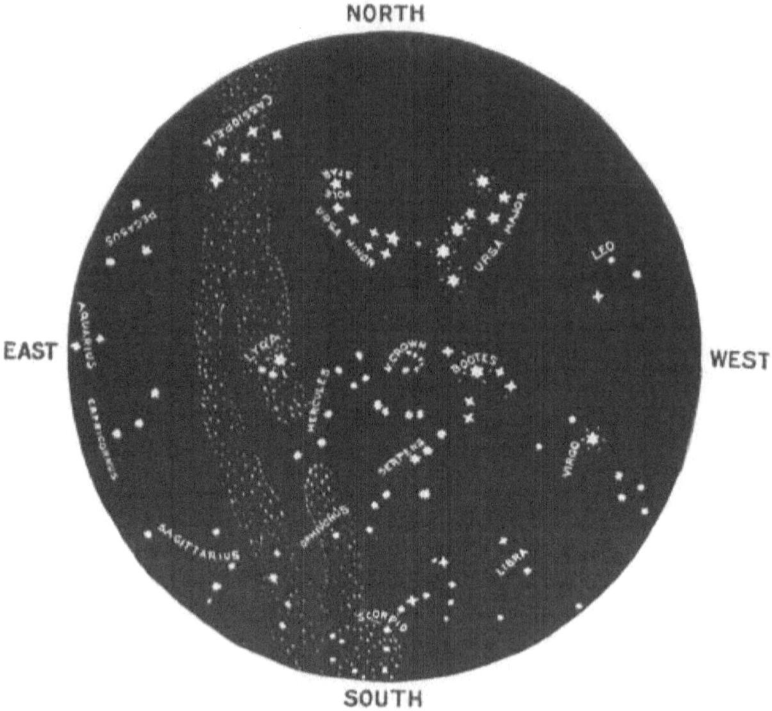

THE SKY IN SPRING

NOTE.—Once upon a time all the educated people spoke Latin. It was the nearest approach to a universal language. So most of the constellations have Latin names. The English, French and German names are all different, but if all children would learn the Latin names they could understand one another. The Spring constellations every child should know are:

LEO, the Lion
LYRA, the Lyre
CASSIOPEIA, the Queen in her Chair
SCORPIO, the Scorpion
URSA MAJOR, the Great Bear
URSA MINOR, the Little Bear
VIRGO, the Virgin

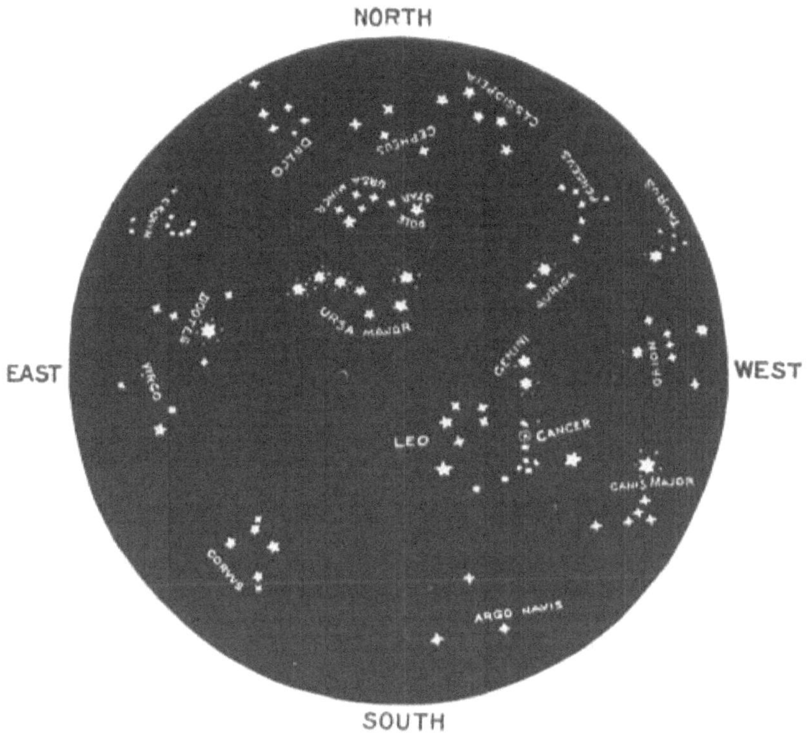

THE SKY IN SUMMER

NOTE.—Every sky map is good for three months, in this way: If this is correct on June 1st at 10 P.M., it will be correct July 1st at 8 P.M., and August 1st at 6 P.M. This is because the stars rise four minutes earlier every night. Thus, after thirty days, any star will rise thirty times four minutes earlier, or 120 minutes, or two hours. Children need not learn all the Summer constellations. The most interesting are:

AURIGA, the Charioteer
CANIS MAJOR, the Big Dog
CYGNUS, the Swan
LYRA, the Lyre
SCORPIO, the Scorpion

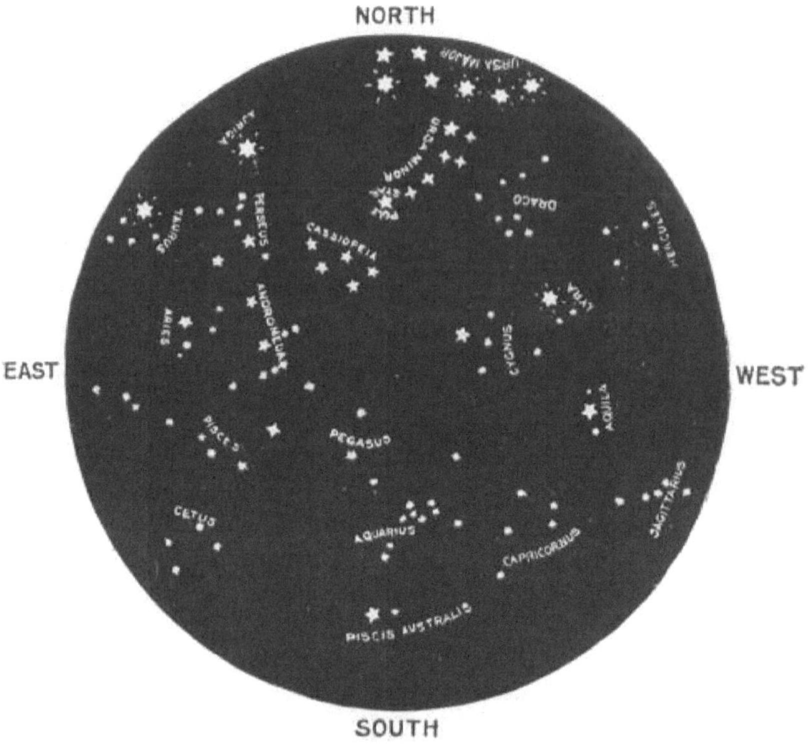

THE SKY IN AUTUMN

NOTE.—This book tells how to find all the most interesting stars and constellations without maps, but many people prefer them. How to use star maps is explained under "The Sky in Winter." The Autumn constellations most interesting to children are:

AQUILA, the Eagle
AURIGA, the Charioteer
CASSIOPEIA, the Queen in Her Chair
CYGNUS, the Swan
LYRA, the Lyre
PERSEUS, Which Has the Arc
TAURUS, the Bull
URSA MAJOR, the Great Bear
URSA MINOR, the Little Bear